This book is to be returned on

MECHANICAL CHARACTERISATION OF LOAD BEARING FIBRE COMPOSITE LAMINATES

Proceedings of the European Mechanics Colloquium 182, 'Mechanical Characterisation of Load Bearing Fibre Composite Laminates', held in Brussels, 29–31 August 1984.

MECHANICAL CHARACTERISATION OF LOAD BEARING FIBRE COMPOSITE LAMINATES

Edited by

A. H. CARDON

Department of Applied Continuum Mechanics,
Free University of Brussels (VUB), Brussels, Belgium

and

G. VERCHERY

Department of Materials, Ecole Nationale Supérieure des Mines,
Saint-Etienne, France

ELSEVIER APPLIED SCIENCE PUBLISHERS
LONDON and NEW YORK

ELSEVIER APPLIED SCIENCE PUBLISHERS LTD
Crown House, Linton Road, Barking, Essex IG11 8JU, England

Sole Distributor in the USA and Canada
ELSEVIER SCIENCE PUBLISHING CO., INC.
52 Vanderbilt Avenue, New York, NY 10017, USA

WITH 18 TABLES AND 182 ILLUSTRATIONS

British Library Cataloguing in Publication Data:

Mechanical Characterisation of Load Bearing
Fibre Composite Laminates (*Conference:
1984: Brussels*)
Mechanical characterisation of load bearing
fibre composite laminates.
1. Fibrous composites 2. Laminated materials
I. Title   II. Cardon, A. H.   III. Verchery, G.
620.1′18   TA418.9.C6

Library of Congress Cataloging in Publication Data

European Mechanics Colloquium (182nd: 1984: Brussels,
Belgium)
Mechanical characterisation of load bearing fibre
composite laminates.

Proceedings of the European Mechanics Colloquium
182...held in Brussels, 29–31 August 1984'.
Bibliography: p.
Includes index.
1. Fibrous composites—Congresses. 2. Laminated
materials—Congresses. I. Cardon, A. H.
II. Verchery, G. III. Title.
TA418.9.C6E97 1984   620.1′18   85-10359

ISBN 0-85334-379-9

The selection and presentation of material and the opinions expressed in this publication are the sole responsibility of the authors concerned.

Printed in Great Britain by Galliard (Printers) Ltd, Great Yarmouth

INTRODUCTION

During the last two decades there has been an increasing interest in lightweight composite materials for applications in the aeronautical, transport and sports industries and in many other fields of mechanical and civil engineering. In order to utilise fully these composite materials in structural applications, it is necessary to have a good understanding of their mechanical characterisation. Thus, starting from the properties of the composite's constituents, their volume fraction and the adhesion between them, modelling theories must be formulated and tested in order to compute the overall mechanical properties of the composite.

If we consider the basic composite material element as a unidirectional long f bre reinforced matrix lamina, then this "material" has an anisotropic character that can generally be reduced to a transverse isotropic constitutive equation with five mechanical constants or functions. Starting from this basic material element and applying the lamination theories, assuming perfect adhesion, then the design of load bearing structures becomes possible.

For polymeric matrix composites, the viscoelastic nature of the matrix introduces a viscoelastic anisotropic behaviour for the composite. This implies time dependent effects and creep, relaxation, frequency, temperature, moisture and other environmental factors influence the mechanical behaviour of the composite. As a consequence, the life-time or durability predictions made from short time tests become a major problem.

In composites subjected to load and various environmental conditions, failures may occur: in the matrix, in the fibres, between matrix and fibres, and between the basic laminae. These different effects are generally covered by the concept of damage. Under constant environmental conditions, three types of failure processes may occur: the instantaneous failure under progressive loading, the creep failure and the fatigue failure. Every failure is a consequence of a certain damage accumulation and it seems very important for practical applications to find a significant macroscopic measure of the "degree" of damage and a relationship between the instantaneous, the creep and the fatigue damage. The influence of the various environmental factors on the damage needs also to be investigated.

The programme of the colloquium included two general invited lectures, one of "Models for analysis of thermomechanical properties and failure of f bre composites and laminates", delivered by Professor Z. HASHIN, and one on "Viscoelastic behaviour and lifetime (durability predictions", given by Professor H.F. BRINSON. The text of Professor Hashin's general lecture is not included in these proceedings, but two principal references could be mentioned: Z. Hashin, "Analysis of composite materials – A survey". Journal of Applied Mechan cs (Transactions of the ASME), Vol. 50, pp. 481–505, September 1983; Z. Hashin, "Analytical methods for fibre composites: Achievements and challenges". Proceedings of the International Symposium "Composites: Materials and Engineering", Center for Composite Mater als, University of Delaware, September 1984.

Three sectional invited lectures concerned three other important aspects:

-   "Damage mechanics and its application to composite materials", by Professor F. SIDOROFF;

-   "Stiffness reduction and development of longitudinal cracks during fatigue loading of cross-ply laminates", by Dr. K.W. SCHULTE;

-   "Adhesion phenomena between phases in composites: the unfolding models", by Professor P.S. THEOCARIS.

Seventeen papers were presented concerning modelling, visco-elastic behaviour and life-time predictions, damage mechanics, impact resistance, stress concentrations and structural applications.

Arising from the colloquium discussions a number of questions were formulated which provide directions for future research in order to arrive at systematic safe design procedures:

1. An important number of failure modes of composites are conse-quences of the structural interactions between the fibre and matrix and between laminae. The question may arise if the classical continuum mechanics description of material behaviour is sufficient for a fibre reinforced composite. It seems that there is a fundamental difference between a homogeneous anisotropic material and a "homogeneised" anisotropic composite, where structural effects between phases play an essential role. Is a structural description and design of a composite material possible?

2. What will be a general macroscopic measure of the degree of damage of a composite material?

3. Is it possible to introduce the concept of damage evolution in the constitutive relations?

4. Is it possible to obtain relations between the instantaneous damage evolution, the creep damage evolution and the fatigue damage evolution?

5. Is it possible to obtain consistent data bases of real-time long term results on a well-defined composite material in order to check prediction methods on the basis of short time tests?

6. What could be a more realistic description of the fibre-matrix adhesion than the concept of perfect adhesion or complete debonding?

Many other questions on basic mechanical description, numerical methods and experimental techniques were touched on during the colloquium.

If some achievements in the description of the mechanical behaviour of fibre composite materials and laminates are real today, a great number of challenges still remain before optimal safe design procedures are established. The end of the story is not for tomorrow, but the challences will hopefully stimulate research for the next decade.

A. H. CARDON
G. VERCHERY

ix

CONTENTS

ACKNOWLEDGMENTS

The colloquium "Mechanical Characterisation of Load Bearing Fibre Composite Laminates" was sposored by the European Mechanics Committee and was held from 29 to 31 August 1984 in the Faculty of Applied Sciences of the Free University of Brussels (V.U.B.).

Financial support was provided by :

- the Belgian National Sciences Foundation (N.F.W.O. - F.N.R.S.)
- the Free University of Brussels (V.U.B.)
- the Belgian Ministry of Education
- the Belgian Organisation for International Cultural Relations
- the USARDS-group (U.K.)
- the Belgian LOTTO
- the A.S.L.K.-C.G.E.R.

The Scientific Committee was composed as followed :

for France   : Prof. G. VERCHERY (co-chairman), Prof. C. BATHIAS,
                Prof. J. PABIOT and Prof. P. HAMELIN

for Belgium : Prof. A. CARDON (co-chairman), Prof. R. DECHAENE,
                Prof. K. GAMSKI and Prof. J. KESTENS

The Belgian Interuniversitary Organising Committee was composed by : A. CARDON (V.U.B.), R. DECHAENE (R.U.G.), K. GAMSKI (U.E.Lg.), J. KESTENS (U.L.B.), M. LEJEUNE (U.C.L.), M. SAVE (F.P.Ms.) and D. VAN GEMERT (K.U.L.).

x
x   x

The local organisation was possible by the participation of all the staff members of the V.U.B.-Composite Coordination Group and more especially A. LESTIBOUDOIS, R. BROUWER, B. NARMON, H. SOL, W. VAN DEN BRANDE en W. DE WILDE.

The most important work in the preparation, during the conference and afterwards was done by the conference secretary Mrs. M. BOURLAU.

A.H. CARDON

Brussels, January 31, 1985.

A. GENERAL AND SECTIONAL LECTURES

VISCOELASTIC BEHAVIOR AND LIFFTIME (DURABILITY) PREDICTIONS

Dr. Hal F. BRINSON, chairman

Virginia Polytechnic Institute and State University, Blacksburg, VA., U.S.A.
Center for Adhesion Science.

ABSTRACT

A method for lifetime or durability predictions for laminated fiber reinforced plastics is given. The procedure is similar to but not the same as
the well known time-temperature-superposition principle for polymers. The
method is better described as an analytical adaptation of time-stress-superposition methods. The analytical constitutive modeling is based upon a
nonlinear viscoelastic constitutive model developed by Schapery. Time dependent failure models are discussed and are related to the constitutive models.
Finally, results of an incremental lamination analysis using the constitutive and failure model are compared to experimental results. Favorable results between theory and predictions are presented using data from creep
tests of about two months duration.

INTRODUCTION

Fiber reinforced plastics (FRP) are being used for many secondary structural applications in automotive, aerospace and other industries. They are
also being considered as primary structural components in the same industries. The reasons for their use focus around weight savings, economy of
construction by tailoring the material to the structural application, improved fatigue tolerance and many other factors. These advantages have
been known for more than a decade and the technology has existed for nearly
as long to design and produce primary structural components utilizing FRP
materials. One might then well ask the question, why are FRP materials not
used even more extensively especially for critical primary structure ?
A major concern has been and remains to be failure modes associated with
the polymer matrix which serves to bind the fibers together and transfer the
load through connections, from fiber to fiber and ply to ply. Some of the
more important matrix dominated failure modes are :

- Interfiber or matrix crazing, yielding, cracking
- Delamination or interply debonding
- Fiber/matrix debonding
- Local fiber buckling
- Fiber pullout
- Matrix failures at connections (bolts, rivets, adhesives).

Prediction of the above failures are complicated by the fact that the polymer matrix is strongly influenced by environmental effects such as temperature, moisture, time, stress, etc...

The matrix is viscoelastic and as a result modulus and strength properties vary with time giving rise to delayed failures or creep ruptures. That is, all of the failures listed above may occur long after the initial design and fabrication process.

From this discussion it is clear that a critical need, before FRP materials can be safely used for a design lifetime of several years, is to develop accelerated testing and analysis procedures such that long term failure predictions can be made from short term test results. In that which follows is the explanation of an accelerated characterization procedure for the prediction of delayed failures which has been developed at VPI & SU by the author and his colleagues. Concepts similar to time-temperature-stress-moisture superposition principles are used in conjunction with laminated plate theory. Because we feel that failures are inherently nonlinear, the testing and analytical modeling for both moduli and strength is based upon nonlinear viscoelastic concepts.

The procedures of necessity include information about the viscoelastic response of the resin and the lamina. Information is presented for two graphite/epoxy systems which are T300/934 and T300/5208. The manner in which both the resin and lamina properties vary with time is presented.

Details of the mathematical nonlinear viscoelastic constitutive model for two and three dimensions are outlined. Time dependent failure models are presented and the need for a time dependent cumulative damage law which must be included in any laminate analysis is illustrated. Potential errors associated with the modeling process are given. Some of these difficulties are determined to be related to the use of a power law approximation for creep.

The results of a laminate testing program as compared with accelerated prediction are given. Test results for 10^4 and 10^5 minutes are given.

Finally, modifications needed to improve the accelerated life prediction techniques are suggested. These include consideration of moisture effects and the reverse thermal effect associated thereto and the development of new time dependent constitutive and failure laws.

EFFECTS OF CURING, AGING AND ENVIRONMENT

How a polymer or an FRP material is manufactured or processed has important applications regarding the long term reliability of a structural component. Further, environmental service conditions are important in ascertaining properties over a design lifetime. For this reason it is appropriate to review a few processing and environmental factors and how they influence performance.

Figure 1 shows a curve representing the free volume as a function of temperature for a polymeric material. It is well known that the slope of the volume-temperature relationship is discontinuous at the glass transition temperature, T_g. In fact, the T_g is often ascertained by slowly increasing or decreasing the temperature and measuring the volume. In many production processes, FRP materials are quenched from the cure temperature (which is usually above the T_g) to room temperature (which is well below the T_g). As indicated in Fig. 1, this procedure potentially leads to a large increase in free volume.

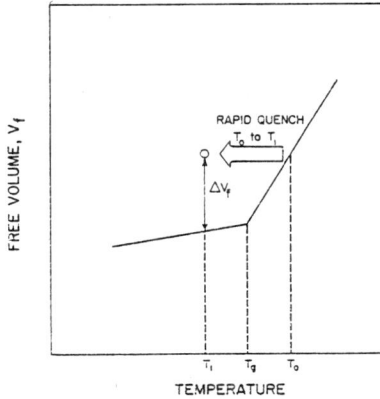

Fig. 1 Relation between free volume and temperature.

As viscoelastic or time dependent mechanical properties can be directly related to molecular processes which are dependent upon free volume, the processing procedures used for composite materials may, in fact, intensify durability concerns. These concerns can often be alleviated by post curing requiring increasing manufacturing time and expense.

Physical aging of polymeric materials gives rise to changes in mechanical properties which are of importance to long term design. These effects are related to but not necessarily the same as the free volume-processing factors discussed above. Figure 2 shows results from Struik (1) regarding the physical aging of PVC. As may be observed, aging strongly affects the creep compliance. Further, Struik showed that aging for PVC was amenable to a shifting procedure to determine master curves similar to the time-temperature-superposition methods and that processes were reversible upon requenching.

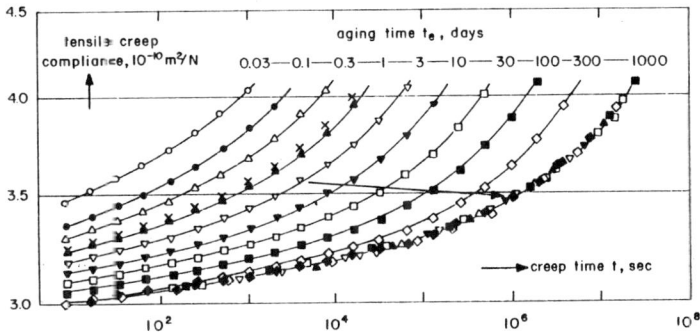

Fig. 2 Results from Struik (1) showing physical aging of creep compliance for PVC quenched from 90 to 20°C. The X's show the results to be thermo-reversible when the specimen was requenched.

It is well known that viscoelastic effects for polymers are increased when the use temperature is at or above the T_g. For this reason, most FRP materials are normally used as structural components only when temperatures are well below the T_g. Unfortunately, the T_g for polymeric materials is a function of their moisture content. Figure 3 shows information given by Whiteside (2) which indicates how the glass transition temperature varies with moisture control for six resin systems commonly used in laminated composites. As may be observed, moisture content can reduce the T_g by as much as 100°C. As a result, the service environment potentially influences the degree and magnitude of viscoelastic and/or durability concerns.

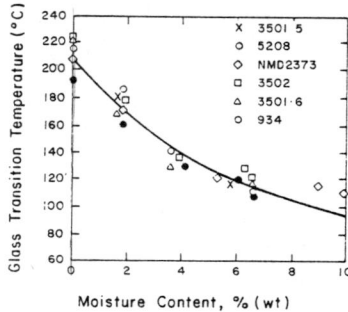

Fig. 3 Effect of moisture content on T_g for six composite resin systems (2).

Moisture effects on composite laminates has been studied extensively (3). The manner by which moisture content changes the T_g was discussed above. However, these and most studies are normally performed under isothermal conditions. For this reason, moisture content is normally assumed to asymptotically reach a saturation level. Recently, Adamson (4) has demonstrated that polymers and composites tend to absorb more water if the temperature is suddenly lowered. This "reverse thermal effect" is shown in Fig. 4 for an AS/3501-5 composite unidirectional laminate. The moisture absorption dramatically increases when the temperature of the immersion bath is changed from 60°C to 25°C even though apparent saturation had been achieved at 60°C. Further, thermal spiking was shown to increase the ability of a material to absorb moisture.

Combinations of the effects shown in Fig. 1-4 tend to indicate some of the underlying reasons for durability concerns for FRP materials used in structural applications. Obviously, a designer and/or a manufacturer would like to have a method of life (durability) prediction which would include these and many other possible processing and environmental effects. In the subsequent pages, a method for lifetime or durability predictions is outlined for FRP materials where temperature and stress are the controlling parameters.
These concepts, we feel, may be extendable to other processing and/or environmental concerns. At least they serve as a point of departure for future discussion and additional research.

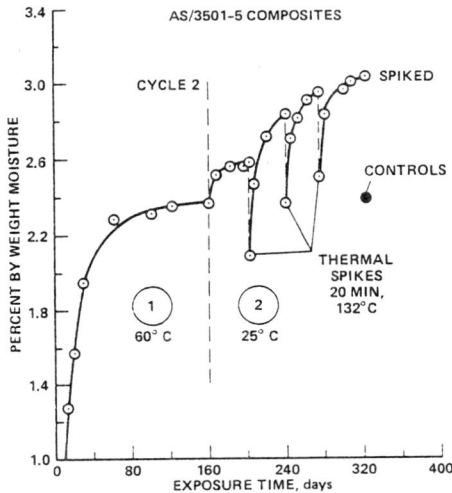

Fig. 4 Reverse thermal effect for AS/3501-5 composites from Adamson (4).

ACCELERATED TESTING AND ANALYSIS PROCEDURES

The ideas to be discussed are those related to the stress analysis of laminated fiber reinforced plastics utilizing the concepts of laminated plate theory. However, the generic ideas could just as easily be applied to micromechanics, finite element or other methods of stress analysis and to other types of composites such as sheet molding compound (SMC) or filament wound pressure vessels.

The first step in our accelerated characterization method is the determination of lamina properties. These are determined using standard testing procedures for tensile specimens (5-7) as illustrated in Fig. 5. That is, tensile tests for the fiber to load angles of 0°, 10° (or 45°), and 90° are performed in a constant rate type test machine. Three stress-strain diagrams are produced as illustrated schematically in Fig. 5b. Each test takes on the order of five to twenty minutes each. A statistical sample of three to five specimens is performed at each fiber angle. Each point on the stress-strain diagram is known as is illustrated in Fig. 5b by the secant modulus. Thus, using the orthotropic transformation equation and an appropriate failure theory, the complete in-plane stress-strain response of a lamina from zero load to failure is known for an arbitrary fiber angle. The variation of the secant modulus with fiber angle for short time (on the order of 10° per minute) including a scatter band is shown in Fig. 6a.

The information so obtained is normally used together with lamination theory or finite element procedures to find the static properties of a general laminate (8). However, such a procedure is insufficient to determine long term (on the order of $10^6 - 10^8$ minutes) design properties which

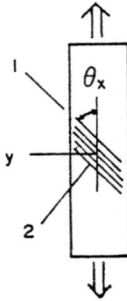

Fig. 5a Lamina tensile specimen   Fig. 5b Stress-strain properties and
         and coordinates.   secant modulus.

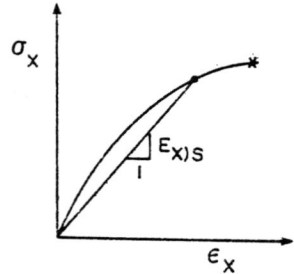

are dominated by viscoelastic, (i.e., creep or relaxation) processes. The
lamina time dependent property surface which is needed to predict long term
design properties of a general laminate is shown in Fig. 6a. This surface
can be produced using an accelerated characterization method based upon a
time-temperature-stress-superposition (TTSSP) concept.

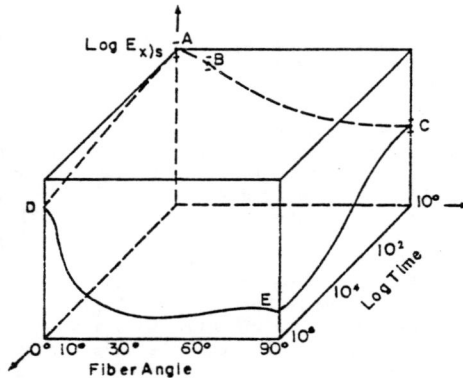

Fig. 6a Long term lamina property surface.

   Fundamentally, the idea is that certain environmental conditions such as
temperature, stress, moisture, aging, etc. serve to accelerate the deformation
process associated with viscoelastic properties. At present only temperature
and stress have been used as accelerating parameters to simplify testing

procedures and data collection efforts. The TTSSP method is illustrated schematically in Fig. 6b. That is, relaxation (creep) testing for 10^c and 90° specimens can be performed at different temperature and stress levels for a short time (on the order of 10^1 - 10^2 minutes) and then shifted to form a modulus (compliance) master curve which would be valid over many decades of time. These master curves together with the orthotropic transformation equation and information on time dependence of [0°] lamina is sufficient to produce the property surface shown in Fig. 6a. In our work on G/E laminates we have assumed the time dependence of [0°] lamina to be negligible.

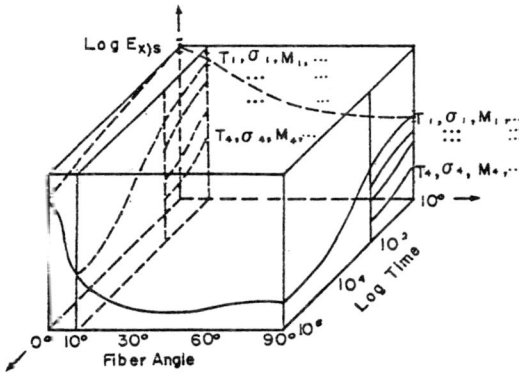

Fig. 6b TTSSP for [10°] and [90°] long term properties.

However, others have shown that fibers or [0°] lamina, including graphite, do have a time dependency which must be known for long time design situations (9,10).

The procedures described above have been included in a lamination theory program which is capable of predicting long term properties of FRP materials from short term data. The following describes some of the analytical methodologies which are input to the laminate analysis.

COMPLIANCE MODEL

The shifting procedures used to determine property surfaces (11) are normally performed graphically and they are both tedious and time consuming. Further, they are semi-empirical at best. The philosophy is based upon the well known time-temperature-superposition principle (TTSP) which is soundly based on molecular mechanisms (12). However, the entire TTSP procedure is strictly applicable at or above the glass transition temperature. Laminated FRP materials are typically used well below the T_g. Use temperatures are still likely below the T_g even when the exacerbating factors of moisture and temperature are included.

A considerable body of literature exists which suggests that shifting procedures analogous to the TTSP are appropriate for the glassy range well below the T_g (13,14). A thorough review of the literature may be found in references (15) and (16).

For the glassy range and for the stress levels often encountered in the matrix of FRP materials, the stress-strain response is likely to be nonlinearly viscoelastic. Further, at or near failure, even for low stress levels, we believe matrix deformation processes are nonlinearly viscoelastic. For these reasons we feel a nonlinear viscoelastic representation of matrix dominated properties of a composite is necessary.

In our earlier work (17-20), a nonlinear power law was used to represent the creep process. While this method can be used to represent creep data, it does not work well for variable stress states and furthermore this approach is basically empirical. More recently, we have used a nonlinear method due to Schapery (16,21-26) which was developed using irreversible thermodynamics.

For the case of uniaxial loading at constant temperature for an isotropic homogeneous nonlinearly viscoelastic material, Schapery's theory reduces to

$$\varepsilon(t) = g_o D_o \sigma + g_1 \int_o^t D'(\psi-\psi') \frac{d(g_2\sigma)}{d\tau} d\tau \qquad (1)$$

where $\varepsilon(t)$ is the time dependent strain, σ is the stress input, D_o is the instantaneous linear compliance, and $D'(t)$ is the transient linear compliance. The quantities ψ and ψ' represent a reduced time scale given by

$$\psi(t) = \int_o^t \frac{dt'}{a_\sigma} \quad ; \quad \psi'(\tau) = \int_o^\tau \frac{dt'}{a_\sigma} \qquad (2)$$

where a_σ is a stress dependent shift factor which is quite analogous to the well known temperature dependent shift factor, a_T, used in the TTSP method. Also, g_o, g_1 and g_2 are nonlinearizing stress dependent parameters.

Equation (1) is a mathematical statement for a time-stress-superposition principle (TSSP). Quite obviously, with an equation of this type it is conceptually possible to have a mathematical framework for including other accelerating parameters such as temperature, moisture, aging, etc. To date, we have only used the equations under isothermal conditions.

Equation (1) is only valid for a uniaxial stress state in an isotropic material. A multi-axial stress state exists in a unidirectional composite laminate for arbitrary fiber angles even if only an axial load is applied to a tensile coupon as shown in Fig. 5a. Schapery's generalization of eq. (1) for an orthotropic material under a multi-axial load state may be written as,

$$\begin{vmatrix} \varepsilon_1(t) \\ \varepsilon_2(t) \\ \gamma_{12}(t) \end{vmatrix} = \begin{vmatrix} D_{11} & D_{12} & 0 \\ D_{12} & D_{22}(t,\tau_{oct}) & 0 \\ 0 & 0 & D_{66}(t,\tau_{oct}) \end{vmatrix} \begin{vmatrix} \sigma_1(t) \\ \sigma_2(t) \\ \tau_{12}(t) \end{vmatrix} \qquad (3)$$

where the average matrix octahedral shear stress has been used to account for stress interaction effects (17,23,25).

The linear viscoelastic compliance, D_{ij}, must be given an analytical form to effectively utilize equations (1) and (3) in a simple manner. In this regard, we have again followed Schapery's suggestion and used a power law to represent the creep process. However, in our opinion, the use of a power law represents the most serious limitation to the utilization of eq. (1) to predict long term data from short term tests results. Tuttle (23) has given an error analysis showing how small errors in the power law parameters as well as the nonlinear parameters g_0, g_1, g_2 and a_σ affect the prediction of long term properties from short term results.

A testing and analysis program was performed by Hiel (16) to determine the viscoelastic response of a T300/934 graphite epoxy composite at various temperatures and stress levels utilizing the procedures discussed above. All testing was for dry materials and both the neat 934 resin and unidirectional T300/934 laminates were evaluated. A few of these results are given in Figs. 7-10. Figures 7 and 8 show the manner in which the creep compliance for the neat 934 epoxy resin and $[90°]_{8s}$ unidirectional T300/934 composite varies with time and temperature. The T_g for the 934 resin is approximately 350°F (177°C) and, as may be observed, viscoelastic effects are quite large for temperatures near the T_g.

The data in Figs. 7 and 8 were shifted to form master curves using the TTSP method (16). Neither the 934 resin nor the $[90°]_{8s}$ laminate appeared to exhibit nonlinear viscoelastic response. That is for the time scale of our tests, the materials appeared to be linear. However, for larger times, nonlinearities are very likely to occur.

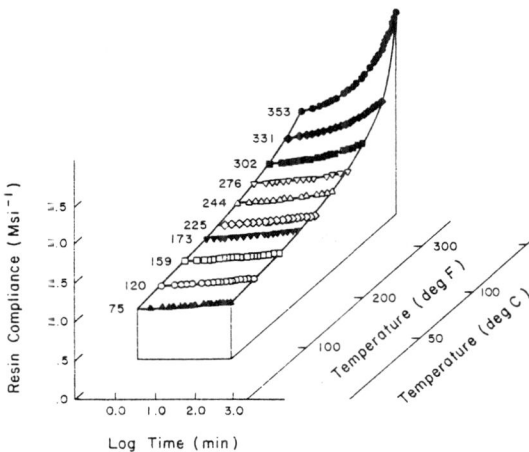

Fig. 7 Temperature dependece of compliance for 934 epoxy resin.

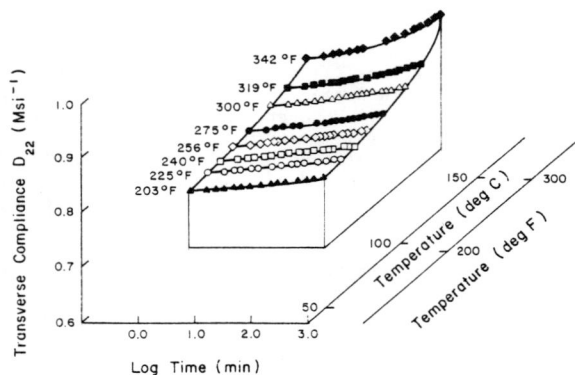

Fig. 8 Transverse or D_{22} compliance of T300/934 as a function of time and temperature.

Figure 9 shows the shear compliance of T300/934 as a function of stress level as determined using a [10°] $_{8s}$ off axis tensile test for a temperature of 320°C (160°C). Tests were performed at many other temperatures with the same general results. Obviously, the T300/934 material exhibits a strong nonlinear response in shear which is very different from the results obtained for the resin or transverse compliance. This result is in keeping with associating viscoelastic deformations with the deviatoric component of stresses. We feel this difference between the shear and transverse compliance behavior may also be related to the manner in which the fiber-matrix interface affects deformation processes in the two situations.

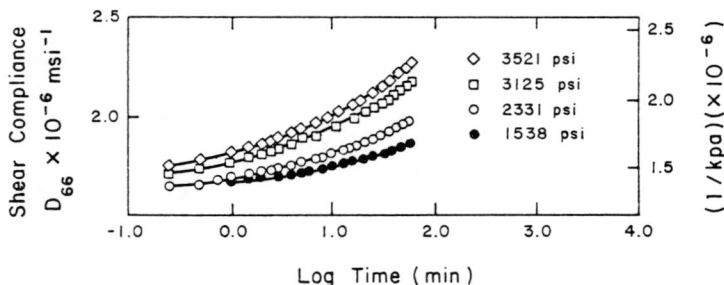

Fig. 9 Shear compliance of T300/934 at 320°C (160°C).

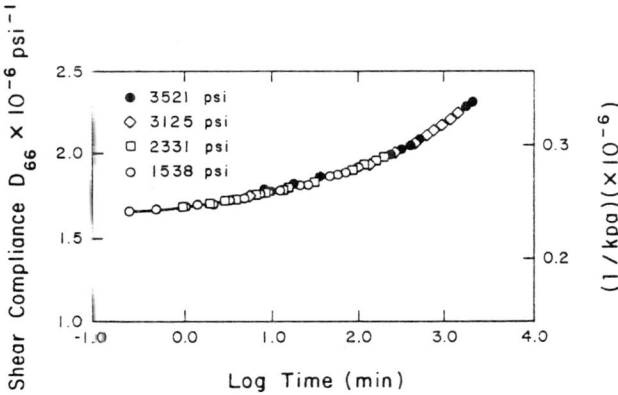

Fig. 10 S_{66} master curve for T300/934 at 320°C (160°C).

The data of Fig. 3 was shifted using a TSSP method with the stress dependent master curve being given in Fig.10. The shifting was done both graphically (symbols) and using the Schapery analysis (solid line) as described
earlier. Quite obviously the Schapery approach we have discussed gives a
good representation of the shifting procedure and can be used as an analytical model of the TSSP method.

FAILURE MODEL

Few failure models are available to quantify time dependent failure processes. The most common form for uniaxial tension assumes that the logarithm
of the time to rupture varies linearly with the applied stress (11,15)

$$\log t_r = A + B\sigma \tag{4}$$

where t_r is the time to rupture under the creep stress σ, and where A and B
are material constants. The functional relation in eq. (4) is often traced
to Zhurkov's modification of the Arrhenius equation in which the parameters
A and B are related to temperature and activation energy. To generalize
this uniaxial form to account for a biaxial stress state in an orthotropic
material, Dillard (17) modified the Tsai-Hill failure criterion to,

$$\frac{\sigma_1^2}{X^2} - \frac{\sigma_1\sigma_2}{X^2} + \frac{\sigma_2^2}{Y^2(t_r)} + \frac{\tau_{12}^2}{S^2(t_r)} = 1 \tag{5}$$

He assumed the strength in the fiber direction, X, to be independent of time,
while the strength in the 90° direction, Y, and the shear strength, S, were
functions of the time to rupture.

Equation (4) is an empirical representation of observed phenomena. Equation (5) is an ad hoc adaptation of an elastic-plastic time independent distortion energy failure law originally developed for isotropic materials. The combination of equations (4) and (5) must be viewed only as a first guess of an appropriate method to calculate dependent failure for composite laminates.

Crochet (26) suggested that time dependent yield stress for uniaxial tension could be expressed as,

$$\sigma = A + B \exp(-c\chi) \tag{6}$$

where A, B and C are material constants. The time dependent parameter χ is related to the difference in viscoelastic and elastic strains by,

$$\chi = [(\varepsilon_{ij}^{V.E.} - \varepsilon_{ij}^{E})(\varepsilon_{ij}^{V.E.} - \varepsilon_{ij}^{E})]^{1/2} \tag{7}$$

Combining eq. (7) and (6) with eq. (1) and assuming a creep power law leads to the result,

$$t_r = [\frac{1}{c\beta\sigma} \ell n \frac{B}{\sigma-A}]^{1/n} \tag{8}$$

where

$$\beta = m \frac{g_1 \cdot g_2}{a_\sigma^n} (1 + 2\upsilon^2)^{1/2}$$

and where the g_1 and g_2 are functions of stress level, a_σ, is the time scale shift parameter, m and n are the power law coefficient and exponent respectively and υ is Poisson's ratio (assumed constant).

The above approach was used by Cartner et al. (21) to represent the delayed creep rupture process in SMC (sheet molding compound). These results are shown reproduced in Fig. 11.

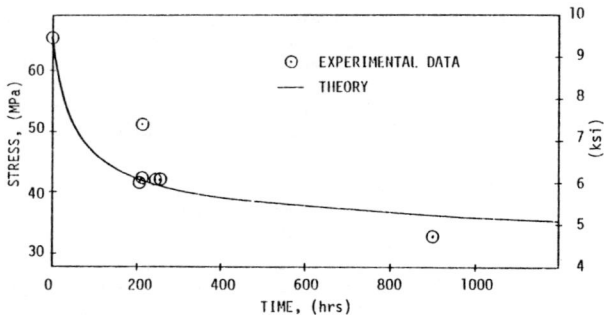

Fig. 11 Delayed failure prediction, SMC-25.

Equation (8) could be included in the modified Tsai-Hill failure criterion given by eq. (5). To date this has not been accomplished but such a procedure has merit and may be useful in predicting some of the matrix dominated failure modes mentioned in the introduction.

Reiner and Weissenberg (27) suggested that failure occurs when the portion of the deviatoric energy stored in a material reaches a characteristic value, called the resilience, which is considered to be a material constant.

This technique has been extensively studied and modified by Brüller (28). Hiel (16) combined this approach with that of Schapery, eq. (1), to obtain the following expression for delayed failure times,

$$t_r = [\frac{1}{\beta^1 \sigma^2} (W - \frac{1}{2} g_o D_o \sigma^2)]^{1/n} \tag{9}$$

where

$$\beta^1 = C \frac{g_1 g_2}{a_\sigma^n} (1 - 2^{n-1})$$

In eq. (9) W is the stored energy and the remaining terms are as defined previously. Hiel used eq. (9) to represent the delayed failure times for a $[90°]_{8s}$ T300/934 composite laminate at 320°C (160°C). The results are shown in Fig. 12.

Again, the Reiner-Weissenberg approach could be included in the modified Tsai-Hill failure criterion given by eq. (5). This has not been accomplished but represents a reasonable approach to time dependent matrix dominated failure mechanisms in FRP materials.

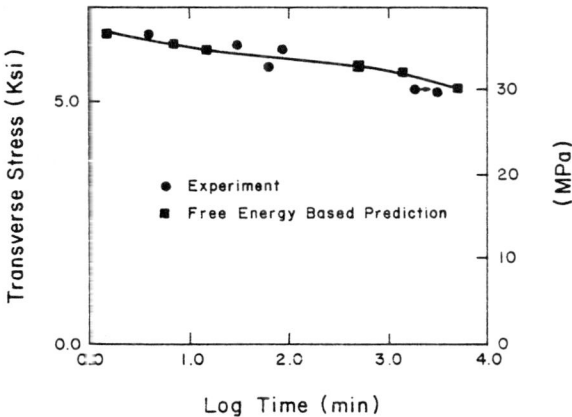

Fig. 12   Comparison of free energy based creep to rupture prediction with experimental results obtained on $[90]_{8s}$ T300/934 laminates at 320°F (160°C).

LAMINATE MODEL

A nonlinear viscoelastic lamination theory program (VISLAP) was developed by Dillard (17) which is capable of predicting long term creep compliance and creep rupture of composite laminates of symmetric layup from short term data. The original program used a nonlinear power law to represent the creep compliances and the modified Zhurkov-Tsai-Hill failure law, eq. (5), to represent creep rupture data. A linear cumulative damage law was used to account for time-varying ply stresses. A complete description of the VISLAP program is given in references (17) and (20).

Dillard conducted an extensive experimental program to establish the creep compliance and the creep rupture response of T300/934 laminates. One hundred and ninety specimens were prepared and tested to obtain creep rupture data from nine different laminates. Delayed failures were produced in all laminates. He then used short term data for T300/934 obtained by Griffith (15) and himself (17) in conjunction with our accelerated charac- terization plan and predicted the creep compliance and creep rupture res- ponse of the T300/934 laminates. Complete details may be found in reference (17). Figure 13 shows the correlation between theory and experiment for a [90°/60°/-60°/90°]$_{2s}$ laminate at 320°C (160°C). In most cases, as may be observed, predictions were reasonable but the approach appeared to be con- servative for creep rupture times in excess of 10^4 minutes.

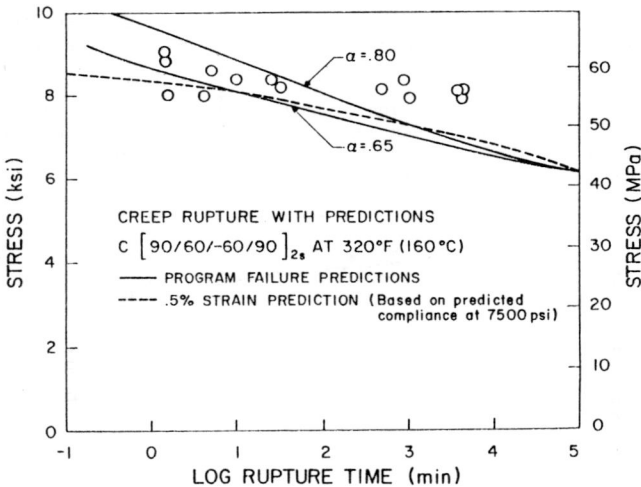

Fig. 13 Creep rupture data with predictions for laminate C [90/60/-60/90]$_{2s}$ at 320°C (160°C).

Our accelerated characterization plan, which has been extensively described elsewhere (5,16,17,18-23), was further validated by Tuttle (23). A new mate- rial which had not been previously subjected to the procedure was selected and studied. The short term (10^2 minutes) creep response of the unidirectional

T300/5208 composite was measured and analytically characterized using the compliance model described earlier. The results were utilized in the VISLAP program to predict the long term ($\sim 10^5$ minutes or ~ 60 days) creep response of two matrix dominated laminates, $[-80/-50/40/-80]_s$ and $[20/50/-40/20]_s$. A comparison between the predictions and the experimental results for the former is shown in Fig. 14. As may be observed, predictions are excellent for times less than 10^4 minutes. For times longer than 10^5 minutes, errors would be expected to increase substantially. However, we feel that the primary difficulty resides with the use of a creep power law representation of the linear viscoelastic compliance.

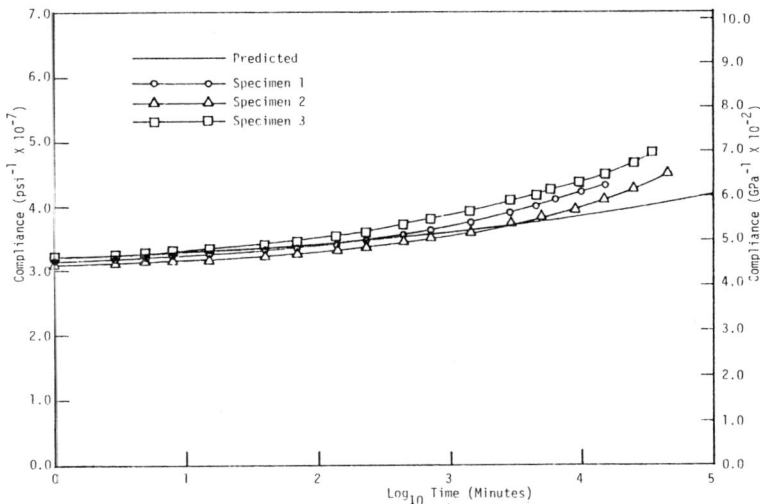

Fig. 14   Comparison between predicted and measured compliances for $[-80/-50/40/-80]_s$ T300/5208 laminate ; creep stress : 76 MPa (11,000 psi), T = 149°C (300°F).

SUMMARY AND CONCLUSIONS

The need for lifetime (durability) predictions using an accelerated characterization procedure was outlined. The cause for such a need was attributed to matrix dominated failure modes which are related to the viscoelastic nature of polymers. Parameters such as free volume, aging, T_g, and moisture were discussed in relation to their effect on viscoelastic properties. Procedures to develop a long term ($10^6 - 10^8$ minutes) property surface from short term (10^2 minutes) data for unidirectional FRP laminates was described. These time-temperature-stress-moisture-superposition (TTSMSP) concepts are similar to but not the same as the well known time-temperature-superposition principle (TTSP).

A nonlinear viscoelastic model due to Schapery was outlined as a convenient method to mathematically represent a time-stress-superposition procedure (TSSP). The procedure was shown to be a valid method to model the stress

18

dependent shear compliance of a T300/934 unidirectional laminate. The 934
resin and the transverse compliance was found to be linearly viscoelastic.
The fiber matrix interface was identified as a possible reason for the discrepancy between transverse and shear compliances.

Several isotropic time dependent failure laws were discussed and their
inclusion in a modified Tsai-Hill failure criterion was indicated. Correlation between theory and experiment was given.

A viscoelastic lamination theory was briefly described. Comparisons between predictions and experiment for matrix dominated laminates were presented. Comparisons were shown to be favorable except for times longer
than 10^4 or 10^5 minutes.

Errors between predictions and experiment can, in general, be attributed
to the many assumptions that were made, e.g., the use of laminated plate
theory, the lamina to be orthotropic and homogeneous, small deformation
theory, etc. However, the most serious difficulties, are likely related
to the power law approximation for creep compliances and the lack of a
rigorous time dependent failure law for laminated fiber reinforced plastics.
All of these features represent new research needs and are currently under
examination.

ACKNOWLEDGEMENTS

The author wish to acknowledge the long standing support by the NASA-Ames
Research Center at Moffet Field, CA, under grants NSG 2038 and NCC 2-71.
The many helpful discussions with Dr. Howard G. Nelson and Mr. Michael
Adamson are greatly appreciated.

Similar studies have been conducted on the nonlinear viscoelastic response of adhesives with support from NASA-Langley Research Center at Hampton,
VA, under contract NAG-1-227 and from the Office of Naval Research, Washington,
DC, under contract N00014-82-K-0185. Appreciation is extended to these
agencies.

The most important acknowledgement goes to the outstanding individuals
with whom I have the privilege to collaborate as a major Ph.D. thesis ad-
visor on topics related to this paper. These are Drs. Y.T. Yeow, W.I. Griffith, D.A. Dillard, C. Hiel and M.E. Tuttle. Related M.S. theses were developed by Messrs. J.S. Cartner, E.S. Caplan and M.A. Rochefort.
Appreciation is extended to each for their timely contributions.

REFERENCES

1. Struik, L.C.E., Physical Aging in Amorphous Polymers and Other Materials,
   Elsevier Scientific Publishing, New York, 1978.

2. Desai, R. and Whiteside, J.B.,"Effects of Moisture on Epoxy Resins and
   Composites", Advanced Composite Materials - Environmental Effects,
   ASTM,STP 658, J.R. Vinson, Ed., pp. 2-20, 1978.

3. Springer, G.S. (ed.), "Environmental Effects on Composite Materials",
   vols. 1 and 2, Technomic Publishing Co., 1981, Westport, CT.

4. Adamson, M.J., "A Conceptual Model of the Thermal-Spike Mechanism in Graphite/Epoxy Laminates", Long-Term Behavior of Composites, STP 813, T.K. O'Brien (ed.), ASTM, Philadelphia, pp. 179-191, 1983.

5. Brinson, H.F., MORRIS, D.H. and Yeow, Y.T., "A New Experimental Method for the Accelerated Characterization of Composite Materials", Sixth International Conference on Experimental Stress Analysis, Munich, September 18-22, 1978.

6. Yeow, Y.T., "The Time-Temperature Behavior of Graphite Epoxy Laminates", Ph.D. Dissertation, VPI & SU, Blacksburg, VA, 1978.

7. Yeow, Y.T. and Brinson, H.F., "A Comparison of Simple Shear Characterization Methods for Composite Laminates", Composites, Jan. 1978, pp. 49-55.

8. Jones, R.M., Mechanics of Composite Materials, McGraw-Hill, New York, 1975.

9. Crossman, F.W. and D.L. Flaggs, "Viscoelastic Analysis of Hygrothermally Altered Laminate Stresses and Dimensions", LMSC-D633086, 1978.

10. Glasser, R.E., Moore, R.C. and Chiao, T.T., "Life Estimation of Aramid/ Epoxy Composites under Sustained Tension", Composites Technology Review, vol. 6, no. 1, Spring, 1984, pp. 26-35.

11. Landel, R.F. and Fedors, R.F., "Rupture of Amorphous Unfilled Polymers", Fracture Processes in Polymeric Solids, (B. Rosen, ed.), Interscience, New York, 1964.

12. Ferry, J.D., Viscoelastic Properties of Polymers, J. Wiley & Son, New York, 1970.

13. McCrum, N.G. and Pogany, G.A., "Time-Temperature-Superposition in the Alpha Region of an Epoxy Resin", J. Macromolecular Science-Physics, B4(1), 1970.

14. Maksimov, R.D. and Urzhumtsov, Y.S., "Predictions of Long Term Durability of Polymer Materials Review", Polymer Mechanics, no. 4, 1977.

15. Griffith W.I., "The Accelerated Characterization of Viscoelastic Composite Materials", Ph.D. Dissertation, VPI & SU, Blacksburg, VA, 1980 ; also Report VPI-E-80-15, 1980, with Morris,D.H. and Brinson, H.F.

16. Hiel, C., "The Nonlinear Viscoelastic Response of Resin Matrix Composite Laminate", Ph.D. Dissertation, The Free University of Brussels (VUB), Brussels, Belgium, 1983 ; also report VPI-E-83-6, March 1983 with A.H. Cardon and H.F. Brinson.

17. Dillard, D.A., "Creep and Creep Rupture of Laminated Graphite/Epoxy Composites", Ph.D. Dissertation, VPI & SU, Blacksburg, VA, 1981 ; also report VPI-E-81-3, March 1981 with H.F. Brinson and D.H. Morris.

18. Brinson, H.F., Dillard, D.A. and Griffith, W.I., "The Viscoelastic Response of a Graphite/Epoxy Laminate", Composite Structures, (I.H. Marshall, ed.), Applied Science, 1981, pp. 285-300.

19. Brinson, H.F. and Dillard, D.A., "The Prediction of Long Term Viscoelastic Properties of Fiber Reinforced Plastics", Progress in Science and Engineering of Composites, (Ed. T. Hayashi et al.), vol. 1, JSCM, 1982, pp. 795-802.

20. Dillard, D.A. and Brinson, H.F., "Predicting Viscoelastic Response and Delayed Failures in General Laminated Composites", ASTM STP 787, Composite Materials : Testing and Design (6th Conference), Dec. 1982, pp. 357-370.

21. Cartner, J.S., Griffith, W.I. and Brinson, H.F., "The Viscoelastic Behavior of Composite Materials for Automotive Applications", Composite Materials in the Automotive Industry, ASME, New York, 1978, pp. 159-169.

22. Hiel, C., Cardon, A.H. and Brinson, H.F., "The Nonlinear Viscoelastic Response of Resin Matrix Composites", Composite Structures, 2, (I.H. Marshall, ed.), Applied Science, 1983, pp. 271-281.

23. Tuttle, M.E., "The Accelerated Viscoelastic Characterization of T300/5208 Graphite/Epoxy Laminates", Ph.D. Dissertation, VPI & SU, Blacksburg, VA, 1984 ; also Report VPI-E-84-9, March, 1984, with H.F. Brinson.

24. Schapery, R.A., "On the Characterization of Non-Linear Viscoelastic Materials", Polymer Engineering and Science, vol. 9, no. 4, 1969.

25. Lou, Y.C. and Schapery, R.A., "Viscoelastic Characterization of a Nonlinear Fiber-Reinforced Plastic, Journal of Composite Materials, vol. 5, 1971.

26. Crochet, M.J., "Symmetric Deformations of Viscoelastic Plastic Cylinders", Journal of Applied Mechanics, vol. 33, 1966, pp. 327-334.

27. Reiner, M. and Weissenberg, K., "A Thermodynamic Theory of the Strength of Materials", Rheology Leaflet, no. 10, 1939, pp. 12-30.

28. Brüller, O.S., "On the Damage Energy of Polymers in Creep", Polymer Engineering Science, vol. 18, no. 1, Jan. 1978, p. 42.

DAMAGE MECHANICS AND ITS APPLICATION TO COMPOSITE MATERIALS

F. SIDOROFF

Ecole Centrale de Lyon, Ecully, France.

SUMMARY

   Damage mechanics has been extensively developed in the last ten years for
lifetime predictions in metal structures.
This paper is a survey of what is damage mechanics and how to develop and
apply its models. Emphasis is laid on a methodological description of the
questions to be asked and the possible ways to answer them. Some examples
are given. As a conclusion the application of these ideas to composite
materials is presented.

INTRODUCTION

   Among the questions which must be answered by the material scientist or
engineer, those related to failure are most important : for any mechanical
structure it is of the utmost importance to know the maximal load it can
bear as well as its lifetime under normal use. Unfortunately mechanical
material modeling, for composites as well as for other materials, is much
more developed towards rheological behaviour than towards failure prediction.
Damage mechanics may be the appropriate approach to deal with this kind of
questions.

   Since the pioneering work of Kachanov in 1958 (1), the damage concept has
been applied to a wide range of engineering problems : creep rupture at
first and more recently lifetime predictions under complex loading, static
failure, crack propagation and bifurcation. However most of these applications have been developed for metals. This paper is a short survey of what
is damage mechanics and how to model, identify and use it. It is not intended
to be an extensive review - such a review can be found for instance in (2) -
but rather as a methodological introduction and guide to the spirit of damage
mechanics and to the kind of informations which can be obtained from it.
Application to composite materials will then be briefly outlined with discussion of some specific aspects.

WHAT IS DAMAGE MECHANICS ?

   Before entering a general discussion of damage mechanics, we shall begin
with two examples showing simple one dimensional damage models.

Kachanov's model for creep rupture
   The concept of damage was first introduced by Kachanov in 1958 (1) for
the description of tertiary creep in metals at high temperature (Fig. 1).

Classical rheology easily accounts for primary and secondary creep. Secondary creep, in particular, is usually described by Norton-Hoff model

$$\dot{\varepsilon} = K \, \sigma^m \tag{1}$$

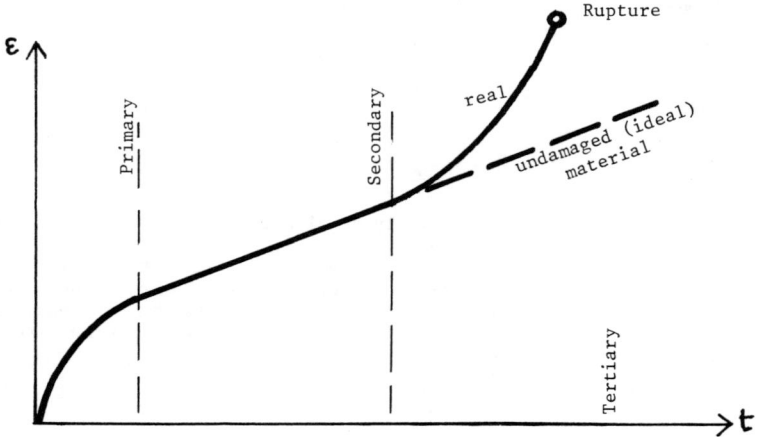

Fig. 1 Creep of metals at high temperature.

In order to describe the degradation process which is responsible for tertiary creep and rupture, Kachanov introduced a damage variable D representing the fraction of destroyed material, thus reducing the effective area of the sample from S to S(1-D). The effective stress, i.e. the effectively applied stress, then becomes

$$\tilde{\sigma} = \frac{F}{\tilde{S}} = \frac{F}{S(1-D)} = \frac{\sigma}{1-D} \tag{2}$$

Kachanov then assumes that the creep law (1) still remains true in tertiary creep provided the effective stress $\tilde{\sigma}$ is used of σ

$$\dot{\varepsilon} = K \, \tilde{\sigma}^m = K \left(\frac{\sigma}{1-D}\right)^m \tag{3}$$

The model is completed by an evolution law giving the damage rate :

$$\dot{D} = H \, \tilde{\sigma}^q = H \left(\frac{\sigma}{1-D}\right)^q \tag{4}$$

In a creep test this equation can be integrated to give

$$D = 1 - [\, 1 - H(q+1) \, \sigma^q \, t \,]^{1/q+1} \tag{5}$$

and failure occurs for D = 1 which is

$$t = t_R = \frac{1}{H(q+1)} \sigma^{-q} \tag{6}$$

Experimental determination of the rupture time t_R as a function of σ then allows to identify the two material constants H and q while K and m are identified as usual from secondary creep data. From this model predicted lifetime can be obtained for complex loading histories.

Brittle elastic model

The behaviour of a brittle elastic material like concrete or rocks in tension can be described, in a first approximation, by the brittle elastic model shown in Fig. 2. Damage resulting from the creation and propagation of a large number of microcracks leads to a decreased elastic stiffness and damage threshold (3). Damage can then be defined from the reduction of stiffness.

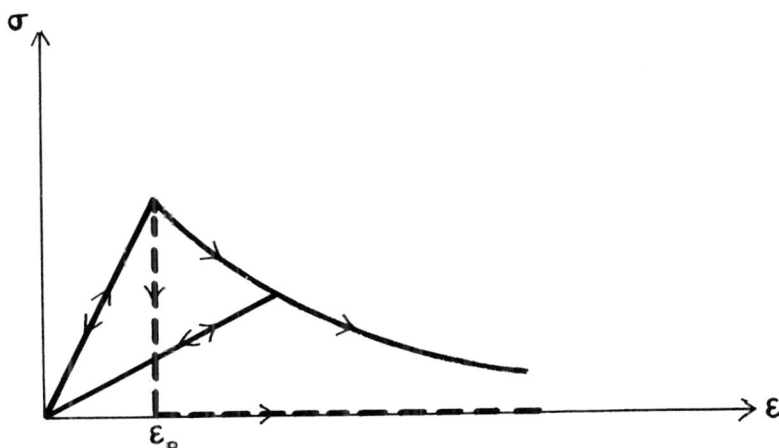

Fig. 2 Brittle elastic model

$$\sigma = \tilde{E} \varepsilon = E_o(1-D) \varepsilon \tag{7}$$

The elastic energy and complementary energy are then given by :

$$W(\varepsilon) = \frac{1}{2} E_o(1-D) \varepsilon^2 = W^*(\sigma) = \frac{1}{2} \frac{\sigma^2}{E_o(1-D)} \tag{8}$$

The energy expanded during damage growth is then

$$\phi = \sigma \dot{\varepsilon} - \dot{W} = G \dot{D}$$

$$G = - \frac{\partial W}{\partial D} = + \frac{\partial W^*}{\partial D} = \frac{1}{2} E_o \varepsilon^2 \tag{9}$$

where the thermodynamic force G can be viewed as an energy release rate. Assuming rate independence and using the general framework of generalized standard materials (4), (or equivalently in this case Hill's maximum power, Ziegler's non linear TIP or Griffith fracture theory), the evolution law for damage is derived from a damage criterion

$$G - G_o \leqslant 0$$

$$\dot{D} = 0 \quad \text{if} \quad G < G_o \quad \text{or} \quad G = G_o, \ \dot{\varepsilon} \quad 0 \tag{10}$$

$$\dot{D} \geqslant 0 \quad \text{if} \quad G = G_o, \ \dot{\varepsilon} \geqslant 0$$

Perfectly brittle elastic materials are obtained for G_o constant ; in this case the material is completely destroyed as soon as the elastic limit ε_o is reached (dashed line in Fig. 2). A more general model is obtained if G_o depends on D and it can be shown that the function $G_o(D)$ may be chosen to fit any $\sigma(\varepsilon)$ curve.

Damage and fracture mechanics

The two examples which have been presented are representative of what is usually called "Damage Mechanics". Generally speaking this title refers to a line of research and a large number of papers and models developed during the last twenty years and which generally can be defined as mechanical and phenomenological models of the material degradation leading to failure and aimed at durability predictions and including mechanical weakening.

Such a definition and a program may as well be applied to another and older field in material sciences : fracture mechanics, but damage mechanics implies an entirely different approach : fracture mechanics deals with one individual, geometrically defined macrodefect. The material is therefore analyzed as a structure. On the contrary, damage mechanics deals with a whole population of microdefects which are statistically described. The material is treated as a continuum.
Fracture mechanics starts from a cracked body. Damage mechanics starts from a perfect, undamaged or virgin material, follows it through its damaged state until it reaches complete damage, which may be interpreted as microscopic failure or instability and corresponds to the appearance of a dominant macroscopic defect.
In fact damage mechanics and fracture mechanics are complementary : damage mechanics or fracture mechanics should be applied according to whether the degradation is a collective or individual mechanism. Strictly speaking, the final stage of damage mechanics should be considered as the initial stage of fracture mechanics, but up to now there does not exist any analysis of this transition. In fact, the local approach to fracture mechanics will be described later on, allows a bypass of the fracture mechanics phase.

HOW TO MODEL DAMAGE

Damage variables

The basic idea behind damage mechanics is the introduction of new damage variables taking into account at the macrocontinuum level, the microscopic degradation process occuring in the material and which finally leads to its failure. These microscopic processes are usually related to the nucleation and propagation of micro voids or microcracks on a sufficiently small scale to be individually ignored and treated as a whole.

In its simplest form, the damage variable may be viewed as a void volume fraction (6) or a destroyed area proportion (1) but these images should not be thought of in a strict sense : damage is a macroscopic internal variable and should be identified from its macroscopic implications rather than from some microscopic measurements.

In most cases a single damage variable D is assumed and it is usually defined from some measurable macroscopic quantity φ which is hoped to be representative of the microscopic degradation process

$$D = \frac{\varphi_o - \varphi}{\varphi_o - \varphi_R}$$

where φ_o and φ_R denotes the values of φ respectively in the initial (and assumed to be perfect) state and at rupture.

When damage results from microvoids, it is natural to use their volume fraction as a damage variable (7) and to obtain it from density changes (8). However, this definition cannot be used when microcracks are dominant and in any case these changes are always very small. Many other quantities φ have been proposed but the most frequently used defines damages as in the elastic brittle model of previous section.

$$D = \frac{E_o - \tilde{E}}{E_o} \qquad \tilde{E} = E_o(1-D) \qquad (11)$$

Damage can then be measured from conventional extensometry (9) or from ultrasonic velocity measurements (10).

Damage may also be defined quite differently (11) for instance as the remaining lifetime under prescribed loading or from micrographic counting of defects, but it should be rememered that damage is a macroscopic internal variable and that any definition can be used as long as it is a good indicator of the degradation mechanism.

Damaged constitutive equations
Any model with internal variables consists in three components :

1) Definition of internal variables, macroscopic and global indicators of the microscopic mechanisms under consideration.

2) Constitutive equations involving the influence of these mechanisms on the behaviour of the material.

3) Evolution equations describing their kinetics as a function of the applied mechanical loading.
Therefore after introducing the damage internal variable, the second step is the formulation of damaged constitutive equations describing the influence of damage on the rheological behaviour.

There are many possible ways of doing this but only one seems systematic enough to be presented in a general framework without reference to any special application : Kachanov's effective stress which has already been mentioned in a previous section. Within this framework, the damaged constitutive equations are obtained by substitution of the effective stress $\tilde{\sigma}$ defined by eq. (2) instead of σ in the undamaged constitutive equation for the virgin material. This postulate has been applied to creep in section 1.1 and its application to the elastic response gives

$$\varepsilon^e = \frac{\tilde{\sigma}}{E_o} = \frac{\sigma}{E_o(1-D)} \qquad \tilde{E} = E_o(1-D) \tag{12}$$

thus leading to the already discussed definition of damage from the elastic stiffness changes. It has also been applied to plastic behaviour but the hardening damage coupling is difficult and not yet fully understood.

Damage evolution law

To complete the model a damage evolution law is needed to follow the damage process under a prescribed loading history and therefore to be able to predict the mechanical weakening of the material as well as its failure which will occur when D reaches 1 or some other critical value D_e. This evolution law will give the rate of damage $dD/d\xi$ and its determination essentially raises two kinds of questions :

a) with respect to which controlling parameter ξ will it be written ?

b) on which variables may it depend ? and how ?

Three types of damages are usually considered.

Creep damage is time controlled. It therefore gives the rate of damage per unit time as a function of stress, damage and some other variables

$$\frac{dD}{dt} = f(\sigma, D, \ldots) \tag{13}$$

Kachanov's evolution equation (4) is the simplest form. The damage rate dD/dt may also depend on the other mechanical internal variables (describing for instance hardening) as well as some external parameters (temperature, environment, etc...).

Rate independent or plastic damage is usually introduced through a damage criterion, like in (10). However, using the standard technique of plasticity (13), it can also be written in incremental form and thus be considered as strain controlled.

$$\frac{dD}{d\varepsilon} = g(\sigma, D, \ldots) \tag{14}$$

or plastic strain controlled. In some cases, it may also be considered as stress controlled but this is not always possible because damage often results in a softening behaviour (13).

In a cyclic fatigue loading relation (14) remains valid and should be integrated over one cycle to give the damage growth per cycle. It is however oftenmore convenient to postulate a fatigue damage evolution law which directly gives the rate of damage per cycle as a function of the cycle parameters

$$\frac{dD}{dN} = h(\sigma_{max}, R, D, \ldots) \qquad R = \frac{\sigma_{min}}{\sigma_{max}} \tag{15}$$

This is a simplified model which neglects the influence of the cycle shape but it is much easier to use and identify in the case of fatigue (9).

These three types of evolution equations should be considered as limiting cases and in many practical situations, they will be simultaneously present. A general evolution equation can be formulated (14) :

$$\frac{dD}{dt} = H(\sigma, \dot{\sigma}, D, \ldots) \tag{16}$$

The general evolution equation can also be postulated by superposition of the elementary ones (13) and (14) or (15). For instance, in the important case of creep fatigue interaction, damage evolution is often assumed as a linear combination of (13) and (15).

$$dD = f(\sigma, D, \ldots) \, dt + g(\sigma_{max}, R, D, \ldots) \, dN \tag{17}$$

Quite a variety of analytical forms for all these functions have been proposed and identified (9),(14). Some results about the micromechanics of void growth can also be used (15).

THREE DIMENSIONAL ASPECTS

All the models which have been discussed until now are simple one dimensional models. In order to use damage mechanics in structure analysis or to include it in finite elementcodes, it is necessary to extend them into complete three-dimensional models and we shall discuss now some questions which arise from this extension.

Isotropic damage
The simplest three dimensional models are obtained by keeping a scalar damage variable D. In thid case, the scalar relation (1) is transformed into

$$\tilde{\underline{\sigma}} = \frac{\sigma}{1-D} \tag{18}$$

defining an effective stress tensor $\tilde{\underline{\sigma}}$ to be substituted for $\underline{\sigma}$ in the undamaged constitutive equation. This means that all stress components are modified in the same way, damage therefore is isotropic. Three dimensional extensions of damage constitutive equations are then easily established. Some care must however be taken to account for the influence of the mean pressure on damage evolution.

For instance, the classical three dimensional extension of Kachanov's model is obtained by assuming instead of (3) and (4)

$$\underline{\sigma} = \frac{3E}{2} \left(\frac{\bar{\sigma}}{1-D}\right)^m \frac{\underline{\sigma}^D}{\bar{\sigma}} \qquad \underline{\sigma} = \frac{1}{3} (\text{tr } \underline{\sigma}) \underline{1} + \underline{\sigma}^D$$

$$\frac{dD}{dt} = \cdots \left(\frac{\hat{\sigma}}{1-D}\right)^q \qquad \bar{\sigma} = \sqrt{\frac{3}{2} \sigma_{ij}^D \sigma_{ij}^D}$$

$$\hat{\sigma} = c \text{ tr } \underline{\sigma} + (1-c) \bar{\sigma}$$

$$\tag{19}$$

where $\bar{\sigma}$ is the usual Von Mises equivalent stress and c is a scalar material constant expressing the influence of the mean stress on $\hat{\sigma}$, that is on damage evolution (16).

Similarly a three dimensional extension of the elastic brittle material model of section 2.2 is obtained by using instead of eq. (7) an isotropic damaged elastic law

$$\underline{\varepsilon} = \frac{1+\nu}{E_o(1-D)} \underline{\sigma} - \frac{\nu}{E_o(1-D)} (tr \underline{\sigma}) \underline{1} \tag{20}$$

and taking an integrated evolution equation giving D as a function of the maximum previously reached value of an equivalent strain (3)

$$D = D(\hat{\varepsilon}_{max}) \qquad \hat{\varepsilon} = \sqrt{<\varepsilon_1>^2 + <\varepsilon_2>^2 + <\varepsilon_3>^2} \tag{21}$$

where <a> denotes the positive part of any scalar quantity a, introducing thus a dissymetry between extension and compression.

Anisotropic damage

Microobservations however show that the microdefects patterns are strongly directional and in many cases isotropic damage cannot be accepted as a reasonable assumption. There are many ways for taking into account damage anisotropy. Most of them rely on the introduction of damage tensor (17). The simplest models are based on a second order symmetric damage tensor D (18),(19) allowing to use instead of eq. (18) a tensorial relationship for instance

$$\tilde{\underline{\sigma}} = \frac{1}{2} [(\underline{1} - \underline{D})^{-1} \underline{\sigma} + \underline{\sigma}(\underline{1} - \underline{D})^{-1}] \tag{22}$$

Such a definition allows a three dimensional extension of one dimensional models. For instance anisotropic extensions of the two simple models of section 2 can be found in (18),(19),(20) and (5). Along a similar line and starting from the definition of damage from elastic stiffness variations, fourth order (21) and eight order (22),(12) damage tensors have also been proposed.
There has been many attempts to define these tensors from some kind of statistical description of the microscopic distribution, configuration and shapes of the microdefects (18), (23), but no one has been really successfull. All these models can only be developed on phenomenological grounds as formal three dimensional extension of existing one dimensional models and there are still many unsolved problems connected in particular with thermodynamic consistency (17), damage evolution law and behaviour dissymetry between extension and compression (5).

Even if these damage tensors have no precise physical meaning, it is however clear that, at best, they are based on some kind of averaging and cannot therefore account for the specific structure and geometry of the actual damage unless it is regular and disordered enough (23),(26). This approach cannot work in case of a highly singular and ordened damage and some other approaches must be developed.
Some have been proposed and we shall come back to them later on about composite materials.

HOW TO USE DAMAGE MECHANICS

Damage cumulation
   The main objective of damage mechanics is to provide a predictive tool for
lifetime evaluation under complex loading. The simplest application is
concerned with damage cumulation that is to situations in which the loading
history consists in a sequence or superposition of simple loading paths.
The example which will be chosen to show the application of damage mechanics
to this kind of problem is the creep fatigue interaction which has been extensively studied for metals at high temperature (9),(14),(27), when submitted
to a succession of cyclic and static loadings. In the simplest case, a cyclic
fatigue loading for N_1 cycles is followed by a static creep loading for time
t_2 or conversely (creep during t_1 followed by N_2 cycles). Miner's linear
cumulation law assumes that in both cases failure will occur for

$$\frac{t_1}{t_R} + \frac{N_2}{N_R} = \frac{t_2}{t_R} + \frac{N_1}{N_R} = 1 \tag{23}$$

where t_R and N_R respectively denotes the time and number of cycles to rupture for a simple creep or fatigue loading at the given level.

   Experimental results do not agree with the simple law : the remaining
lifetime t_2 after N_1 cycles is greater than expected from (23) while the
converse is true when fatigue follows a creep loading

$$\frac{t_1}{t_R} + \frac{N_2}{N_R} < 1 \qquad \frac{t_2}{t_R} + \frac{N_1}{N_R} > 1 \tag{24}$$

   In terms of damage, Miner's law can be interpreted as resulting from damage independent damage evolution laws

$$\frac{dD}{dt} = \frac{1}{t_R(\sigma)} \qquad \frac{dD}{dN} = \frac{1}{N_R(\sigma_{max}, R)}$$

Non linear damage cumulation is therefore obtained from the usual damage
evolution laws as described in section 3.3. Satisfactory results are
usually obtained from eq. (17), ref. (28). This is clearly understood in
Fig. 3 showing the evolution of damage in creep, fatigue and complex paths.

Lifetime predictions in structures
   In real structures lifetime predictions should be included as a part of
structure analysis. Therefore damage mechanics is to be included in the
finite element codes currently used for structural analysis and design.
Such computations have been performed and they can be divided into two
types.

   The first type of application may be called "uncoupled damage" analysis
because it neglects the influence of damage on the constitutive equations :
A classical structural analysis is performed with the usual (undamaged)
constitutive equations (either elastoplastic, viscoplastic or viscoelastic).
The damage evolution law is then written at each point and integrated starting from the computed stress state. Lifetime and the location of rupture
can then be determined by looking for the first time and place where D

reaches 1.

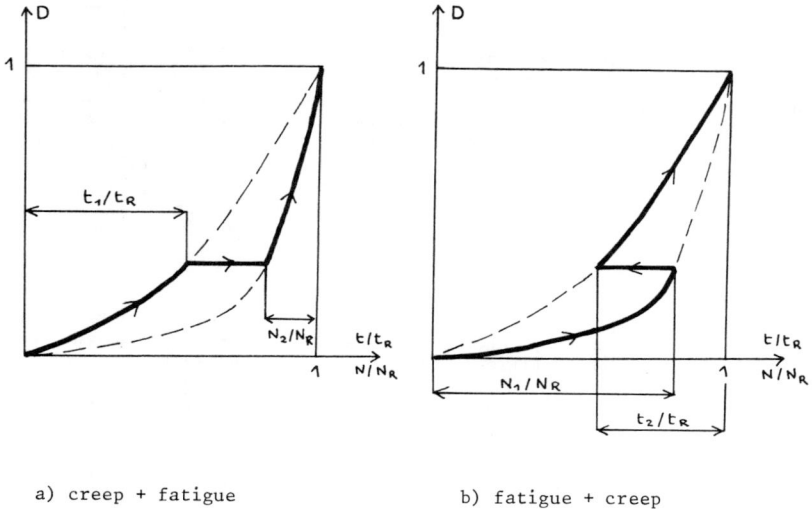

a) creep + fatigue   b) fatigue + creep

Fig. 3 Creep-fatigue interaction.

Such an analysis is obviously a crude approximation since it does not take into account the mechanical weakening of the material before rupture. It is however easy to handle since it can use an existing finite element code and only requires a supplementary integration of ordinary differential equations. This approximation may be justified by the fact that there is usually a steep damage growth before failure so that the period during which the influence of damage on the stress distribution may be significant usually is small compared with the total lifetime.

Furthermore this is a conservative assumption because mechanical weakening usually leads to load transfers from the critical point of the structure to other parts. Such a computation has been for instance performed for lifetime predictions in turbine blades (28).

Local approach to fracture mechanics

The second type of computation explicitly takes care of this coupling. The whole set of constitutive equations including damage effect and damage evolution must then be included in the numerical solution scheme. This requires the development of new finite element codes, including damage, and leads to some numerical problems which are not yet fully understood. This kind of analysis therefore implies heavy computational work and it has only been performed for very simple models (elasticity or creep, iso- tropic damage).

Such an approach however is very powerful : since it takes into account the mechanical weakening of the material, it is able to describe the load transfers resulting from it and the subsequent development of the damage zones.

It can therefore account for damage propagation that is to some kind of crack propagation, the crack being defined as the set of all points where D = 1 and which are therefore without any mechanical strength left. Thus damage mechanics used in this way provides a local approach to fracture mechanics.

Until now, it has only been applied to simple models but these applications deal with problems which are usually very difficult to handle within the framework of classical fracture mechanics : crack bifurcation (29) and crack initiation and multiplication in a reinforced concrete beam (3) for instance. Such computational methods will also be necessary for the analysis of the experiments which are required for the identification of damaged constitutive equations and evolution laws : since damage strongly depends on the mean tension, as mentioned earlier, the classical experimental tests, with uniform stress and strain states, will not be sufficient for the complete identification of the model : it will be necessary to use more complex tests which will require a comparison with numerical results.

In the case of perfectly brittle elastic materials, which has been described in section 2.2, some theoretical and analytical results have also been obtained and variational methods have been developed (31). In particular, a complete theoretical analysis of the stationary propagation of a rectilinear crack in plane elasticity is available (32).

WHAT ABOUT COMPOSITES ?

Unidirectional damage
The development and application of damage mechanics which have been described earlier mostly refer to metals and sometimes concrete. As far as composite materials are concerned, there exists a number of papers referring, more or less explicitly, to this concept but these works remain isolated and there is no systematic attempt to develop damage mechanics in the scope of composite materials. Yet, in some respects, fiber reinforced composite materials fit quite well within the range of application of damage mechanics.

Indeed a bundle of elastic fibers is the simplest mechanical model for damaging materials and it is often used as an introduction to Kachanov's idea of damage and effective stress. More precisely, this bundle can exactly be described by the brittle elastic model of section 2.2, the damage evolution law resulting from the statistical distribution of failure strain ε_f. In particular the perfectly brittle elastic material corresponds to the case where all fibers have the same failure strain. More generally, the progressive degradation process which underlies the damage concept is well recognized for composite materials. Extensive experimental studies using quite a number of different techniques have been performed to characterize and identify it. The stiffness reduction which is commonly used for the definition of damage also is a well established experimental fact for fiber reinforced composites, much better established in fact than for metals in which it is extensively used.

Therefore a unidirectional fiber reinforced composite material loaded in tension appears as the ideal application for the one dimensional damage models as they have been described in section 2 and 3. This kind of application does not raise any new conceptual problems with respect to what has been developed earlier for other materials. Very few attempts have been made (33) but a wide application field is open, in particular in connexion with fatigue damage.

General damage models

These one dimensional scalar models however will not be sufficient to be used for lifetime predictions in composite structures which is the ultimate goal of damage mechanics. Even in the simplest case of a multilayer laminate plate, a complete model, including three dimensional aspects will be required. Some specific aspects about damage in composite materials will have to be taken into account for the development of such a model and this will require significant deviations for the usual concepts and model now being used for metals and concrete.

One of these difficulties is related to the fact that in a composite material, the basic undamaged behaviour is viscoelastic and therefore usually described through hereditary integral laws. Such a constitutive equation is not easily coupled with damage which essentially is an internal variable approach. Furthermore damage destroys linearity so that the usual techniques based on Laplace transforms can no longer be used. For instance, the simple introduction of Kachanov's effective stress in a viscoelastic constitutive equation will result in a highly sophisticated integro-differential system. More generally, such questions as the coupling of damage with complex moduli and frequence analysis, the definition of damage from viscoelastic properties, the influence of temperature or moisture do raise difficult conceptual problems.

Another problem is related to damage definition : many mechanisms are known to be responsible for composite damage (fiber failure, matrix cracking, interface decohesion,...) and their interaction is not clearly understood. Averaging them in a single damage variable cannot be realistic and several damage variables will be needed. They are also strongly related with the geometrical and mechanical ansitropy of composite materials so that the problems discussed at the end of senction 4.2 cannot be avoided. The proposed anisotropic damage tensors are inappropriate and some other kind of damage description must be thought of. For instance, several scalar variables may be introduced (24),(25) therefore accounting for several damage mechanisms. Directional damage definition like in (26) can also provide a reasonable approach. In any case, some physical and statistical insight in the existing damage patterns should be useful.

At last, interface decohesion and the resulting friction may result in unusual damage influence on the behaviour (34),(35).

CONCLUSION

Application of damage mechanics to composite materials is an exciting and promising challenge for the next few years. Much material has been gathered about metals which can certainly be used as a starting point but the specificity of composite materials will require strong changes. However these developments may rely on a large amount of experimental results and quite a variety of characterization techniques : damage has been extensively studied in composite materials. Furthermore many of these damage mechanisms can be interpreted on a micromechanical basis, thus providing answers to the physical questions raised in the three steps mentioned in section 3.2. Within this framework homogenization techniques should play an important part in these developments because they can predict the homogenized mechanical behaviour resulting from a given damage pattern and the evolution of this pattern by introducing some kind of microfracture mechanics in the local stress analysis.

REFERENCES

1. Kachanov, L.M., "Time to failure under creep conditions", Ak. Nauk. SSR. Otd. Tekh. Nauk. n°8, 1958, pp. 26-31.

2. Lemaitre, J., "Damage modelling for prediction of plastic or creep fatigue failure in structure". SMIRT 5, 1979.

3. Mazars, J. and Lemaitre J., "Application de la théorie de l'endommagement au comportement non linéaire et à la rupture du béton de structure". Annales de l'ITBTP, n°491, 1982, pp. 114-138.

4. Halphen, B. and Nguyen, Q.S., "Sur les matériaux standard généralisés". J. Mécanique, vol. 14, 1975, pp. 39-64.

5. Supartono, F. and Sidoroff, F., "Anisotropic damage modelling for brittle elastic materials", 5th Polish-French Symposium "Non linear problems of mechanics". Rydzyna, 1984, to appear in Arch. Mech.

6. Perzyna, "The onset of instability and fracture as two particular stages of inelastic flow process". 5th Polish-French Symposium "Non linear problems of mechanics", Rydzyna, 1984, to appear in Arch. Mech.

7. SCHMITT, J.H. and Jalinier, J.M., "Damage in sheet metal forming". Part 1 - Physical behaviour. Acta Met. Vol. 30, 1982, pp. 1789-1798.

8. Baudelet, B. and Jaliner, J.M., "Endommagement. Aspects physiques, mécaniques et industriels". 5ième Congrès Français de Mécanique, Marseille, 1981.

9. Lemaitre, J. and Chaboche, J.L., "Aspects phénoménologiques de la rupture par endommagement". J. Mec. Appl. vol. 2, 1982, pp. 317-365.

10. Bamberger, Y., Cannard, G. and Marigo, J.J.,"Microfissuration du béton et propagation d'ondes ultrasonores". Mech. Behav. of Anisotropic Solids, Ed. J.P. Boheler, CNRS et Martinus Nijhoff, 1982, pp. 715-734.

11. Chaboche, J.L., "Lifetime predictions and cumulative damage under high temperature conditions". ONERA - TP. 1980, 120 p.

12. Chaboche, J.L., "Sur l'utilisation des variables d'état interne pour la description du comportement viscoplastique et de la rupture par endommagement". 3rd Polish-French Symposium "Non linear problems of mechanics", Cracovie, 1977, ed. W.K. Nowacki, PWN, 1980, pp. 137-160.

13. Sidoroff, F., "On the formulation of plasticity and viscoplasticity with internal variables". Arch. of Mech., vol. 27, 1975, pp. 807-819.

14. Ostergren, W.J. and Krempl, E., "A uniaxial damage accumulation law for time varying loading including creep fatigue interaction". J. Press. Vess. Tech., vol. 101, 1979, pp. 118-124.

15. Beremin, F.M., "Experimental and numerical study of the different stages in ductile rupture – Application to crack initation and stable crack growth". North Holland, 1981, pp. 185-205.

16. Hayhurst, D.R. and Leckie, F.A., "The effect of creep constitutive and damage relationships upon the rupture time of a solid circular torsion bar". J. Mech. Phys. Solids, vol. 21, 1973, pp. 431-446.

17. Sidoroff, F., "Description of anisotropic damage – Application to elasticity – Physical non linearities in structural analysis". Ed. J. Hult, J. Lemaitre, Springer, 1981, pp. 237-244.

18. Murakami, S. and Ohno, N., "Creep damage analysis in thin walled tubes – Inelastic behaviour of pressure vessels and piping components". A.S.M.E., New York, 1978, pp. 55-69.

19. Cordebois, J.P. and Sidoroff, F., "Damage induced elastic anisotropy". Mechanical behaviour of anisotropic solids, Ed. J.P. Boehler, C.N.R.S., Martinus Nijhoff, 1982, pp. 761-744.

20. Cordebois, J.P. and Sidoroff, F., "Endommagement anisotrope en élasticité et en plasticité". J. Mécan. Théo. et Appl., n° spécial 1982, pp. 45-60.

21. Chaboche, J.L., "Le concept de contrainte effective appliqué à l'élasticité et à la viscoplasticité en présence d'un endommagement anisotrope". Mechanical behaviour of anisotropic solids, Ed. J.P. Boehler, C.N.R.S. and Martinus Nijhoff, 1982, pp. 737-760.

22. Fitzgerald, J.E., "Some divagations with respect to an operational definition of damage". Workshop on a Cont. Mech. approach to damage and life-prediction., Carollton, 1980, pp. 153-158.

23. Leckie, F.A. and Onat, E.T., "Tensorial nature of damage measuring internal variables". Physical non linearities in structural analysis. Ed. J. Hult et J. Lemaitre, Springer, 1981, pp. 140-155.

24. Krajcinovic, D., "Constitutive equations for damaging materials". J. Appl. Mech., vol. 50, 1983, pp. 355-360.

25. Poss, M., Proslier, L. and Ladeveze, P., "Endommagement et rupture des matériaux composites tridimensionnels". J.N.C. 3, Paris, 1982.

26. Ladeveze, P., "Sur une théorie de l'endommagement anisotrope". Failure criteria of structured media, Villard de Lans, 1983.

27. Chrzanowski, M., "Use of the damage concept in describing creep fatigue interaction under prescribed stress". Int. J. Mech. Sc., vol. 18, 1976, pp. 69-73.

28. Chaboche, J.L. and Stolz, C., "Détermination des durées de vie des aubes de turbines à gaz". Revue Française de Méc., vol. 52, 1974, pp. 37-47.

29. Benallal, A., Billardon, R. and Lemaitre, J., "Failure analysis of structures by continuum damage mechanics". I.C.F.G., New Dehli, 1984.

30. Benallal, A. and Marquis, D., "Multiaxial low cycle fatigue prediction by damage mechanics". A.S.T.M. STP 1983.

31. Bui, H.D., Dang Van, K. and De Langre, E., "A local approach to crack growth in creeping material by a shear DCB model". I.C.F.G., New Dehli, 1984.

32. Bui, H.D. and Ehrlacher, A., "Propagation dynamique d'un zone endommagée dans ur solide élastique fragile en mode III en régime permanent". CRAS, Paris, ¬.290.B, 1980, p. 273.

33. Poursatip, A., Ashby, M.F. and Beaumont, P.W.R., "Damage accumulation during fatigue of composites". Script. Metall., vol. 16, 1982, pp. 601-606.

34. Chenard, J.F. and Sidoroff, F., "Sur la plasticité induite par endommagement dans les matériaux composites". Journées AMAC, Lyon, Décembre 1983.

35. Arnould, J.F., "Damage of laminated composites". 5th Polish-French Symposium "Non linear problems of mechanics", Rydzyna, June 1984.

STIFFNESS REDUCTION AND DEVELOPMENT OF LONGITUDINAL CRACKS DURING FATIGUE
LOADING OF COMPOSITE LAMINATES

K. SCHULTE

DFVLR, Institut für Werkstoff-Forschung, Köln, West-Germany.

INTRODUCTION

Favourable mechanical properties of carbon-fiber reinforced plastics as
a high elastic modulus and high fracture strength combined with low specific
weight make them more and more attractive as a construction alloy in many
fields of engineering purposes. In spite of the fact that a great body of
literature is available about the mechanical properties of composite materials, the understanding of the fatigue and fracture behaviour is not yet
satisfactory (1). Composite materials have, especially in tension-tension
fatigue (this type of load will be discussed in this paper), a superior
fatigue behaviour compared to metals. Even though they fail during fatigue
loading. Whereas in long-time fatigue the maximum stress in the load cycle
is about 30 % of the ultimate stress in metals, in composite materials a
maximum stress in a load cycle can by all means exceed 60 to 80 % of the
ultimate stress.

Stiffness reduction, continuously registered during fatigue loading turned
out to be an excellent analogue to describe damage development during fatigue.
The various damage mechanisms leading to stiffness reduction will be presented. Especially the development of longitudinal cracks and the rupture of
the load bearing fibers in the 0°-direction will be discussed.

Carbon fibers itself are nearly totally insensitive to fatigue loading (2),
nevertheless failure of composite materials does occur during tension-
tension fatigue. Therefore, it is even more questionable what the reasons
are for the fracture of 0°-fibers.

Fiber fracture could be explained by means of micromechanisms, however,
the response of the fibers itself to changes of the micromechanical stress-
state has not yet been discussed satisfactorily. This will be tried in the
present paper in a separate chapter.

TEST PROGRAM

Composite materials have the advantage that it is possible to design the
alloy appropriate to the expected loads, that means to optimize the alloy
with respect to the number of plies and their orientation. For this investigation a cross ply laminate with fibers in the 0° and 90° direction was
chosen.

The alloy system used was the T300 fiber (Normco) with a 914C matrix system (Ciba-Geigy). The specimen used had dimensions of 200 mm x 25.4 mm.
The thickness was about 1.9 mm, as a result of 16 plies used for each

laminate type.

Tension-tension fatigue tests were performed with an R-ratio ($R = q_u / \sigma_o$) of 0.1. The load ratio was chosen in such a way that the expected life of the specimen was in the range of 10^5 to 10^6 load cycles.

During the fatigue tests a change of the secant of the elastic module was continuously measured. To characterize the damage development in the specimen, different techniques, as there are the replication technique, light and scanning electron microscopy (SEM) combined with the deply technique as well as the X-ray technique were used (3).

EXPERIMENTAL RESULTS AND DISCUSSION

It was shown in previous publications that the stiffness change (secant of elastic module) is an excellent parameter to describe the damage development occurring in composite materials during fatigue loading (4-6). The change of the secant of the elastic module is a well defined engineering constant. Its change can directly be related to stress redistributions which can be expected, if internal damage occurs in composite laminates. A typical stiffness reduction curve of the cross ply laminate is shown in Fig. 1.

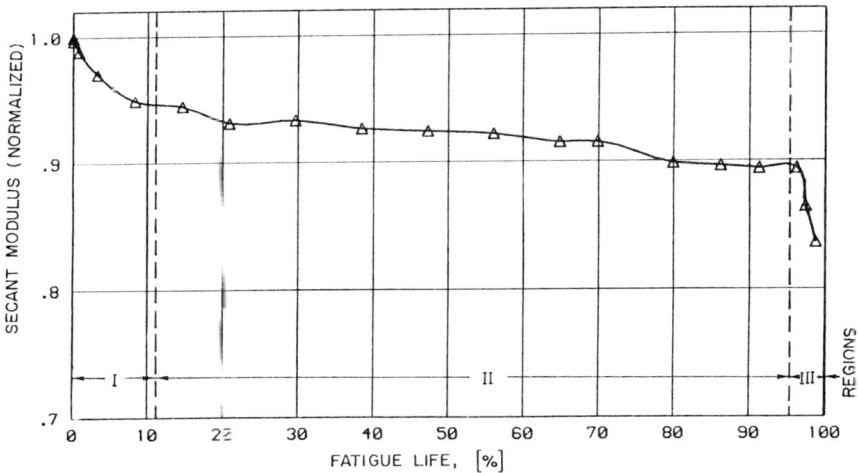

Fig. 1 Typical stiffness reduction curve for a $[0,90,0,90]_{2s}$-laminate.

The observed laminate types have characteristic stiffness reduction curves. The shape of the curves suggests three regions of interest : An initial region (stage I) with a rapid stiffness reduction of 2-5 %, an intermediate region (stage II), in which an additional 1-5 % stiffness reduction occurs in an approximately linear fashion, and a final region (stage III), in which stiffness reduction occurs in abrupt steps ending in specimen fracture.

a) Stage I

During fatigue loading in stage I the formation of transverse cracks
parallel to the fibers can be observed very early. They develop in those
plies which are perpendicular or in some angle to the load axis. For every
type of laminate there is a constant distance between the transverse cracks.
This has been called in the literature the characteristic damage state (CDS),
(7,8). In Fig. 2 is shown a penetrant enhanced X-ray radiograph showing
the fully developed CDS in a cross ply specimen. A penetrant (ZnI_2) was
used to make the transverse cracks visible. In a fatigue test the constant
distance between the transverse cracks of the CDS is already developed after
only a few hundred load cycles (4). That means, the development of transverse cracks dominates the stiffness reduction ascertained in early fatigue
life (9-12).
An analytical description of the failure strength of cross ply laminates
with transverse cracks was given recently (13).

Fig. 2   Penetrant enhanced X-ray radiograph.   Transverse cracks developed
during formation of the CDS.

b) Stage II

During fatigue in stage II a constant decrease of the elastic module
can be observed, visible in the linear part of the curves shown in Fig. 1.
The damage mechanisms occuring in stage II seem to be the typical fatigue
mechanics, whereas the damage mechanisms which can be observed during
stage I, do also occur in static tests. A typical mechanism to be observed
during stage II is the development of edge delaminations which are the main

influencing mechanisms for the reduction of the elastic module (14). In Fig. 3 is shown in a penetrant enhanced X-ray radiograph the delamination starting at the edge of the specimen produced during fatigue loading.

Fig. 3 Penetrant enhanced X-ray radiograph. Transverse and longitudinal matrix cracks and delamination growth.

Besides delamination Fig. 3 shows the transverse cracks of the CDS and additional longitudinal cracks along the 0°-fibers which have developed from the transverse cracks. These longitudinal cracks increase in length and density with increasing number of fatigue cycles (9). In Fig. 4 is explained the development of these longitudinal cracks. If a specimen is loaded into tension, the 0°-plies carry most of the load. Because of internal damage (as there are the CDS and delaminations) an additional increase in the load of the 0°-fibers occurs. Because of the Poisson strain a contraction within the 0°-plies occurs. This contraction is hindered by the 90°-plies, which have a higher elastic module than the 0°-plies in the Y-direction (the width direction of the specimen). In the 0°-plies develop stresses in the Y-direction (here called σ_2), which can lead especially under a fatigue load to longitudinal cracks. Starting of the cracks and their development into the longitudinal direction can now be explained as a typical fatigue mechanism because the σ_2 stresses might not be high enough to develop cracks under static loading, but they are high enough to develop cracks under fatigue loading.

In the following it will be tried to describe the influence of the formation of longitudinal cracks to stiffness reduction. We consider an equal load in each consecutive load cycle. Because of the Poisson ratio we have a constraint effect during loading, schematically shown in Fig. 5a.

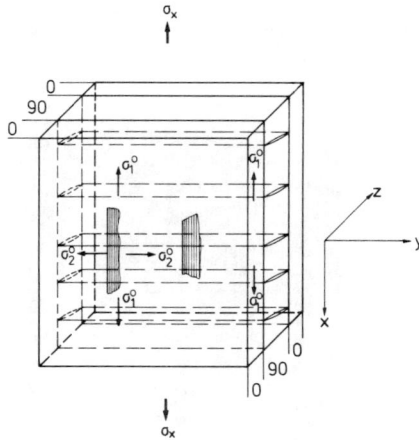

Fig. 4 Schematic presentation of the stress distribution in a [0,90,0]-laminate with transverse cracks and development of longitudinal cracks.

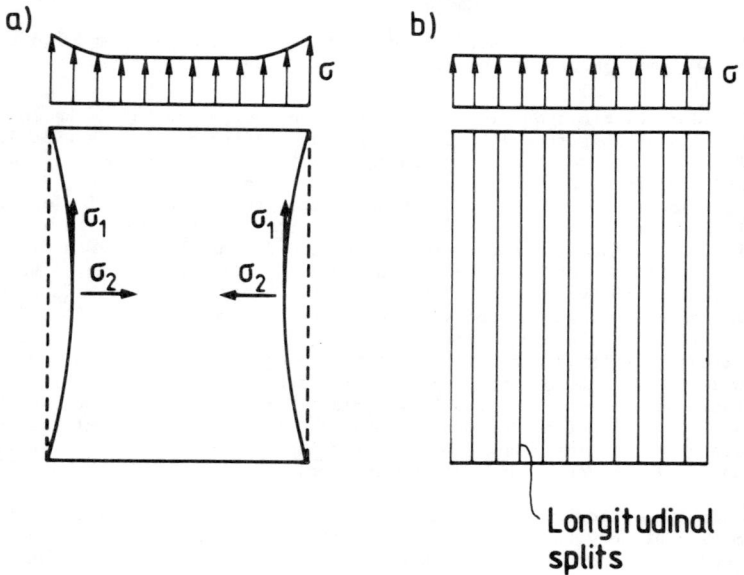

Fig. 5 Stress distribution and schematic presentation of the constraint effect in a unidirectional ply : (a) undisturbed – (b) ply with longitudinal cracks.

In a specimen with no longitudinal cracks the fibers near the edge are bowed, while in the center of the specimen they remain straight, which also means that the stresses of the fibers at the edge are higher than in the interior. The higher stresses near the edge do not contribute to an increase in displacements (length) of the specimen. They just are necessary, because of constraining the near the edge fibers have to be longer, as those in the interior. However, we also have to consider equal displacements over the whole cross section of the specimen.

If now longitudinal cracks develop (Fig. 5b), the zero degree ply consists of a number of parallel fiber bundles, where each bundle is loaded separately. The constraining effect in each bundle is quite small and can be neglected. In this case the σ_2 stresses nearly dissapear. We now reach an equal stress distribution, but the stresses in the interior are higher and those near the edge lower as in the case where no cracks have developed. The higher stresses lead to an incrase in length of the fibers and therefore of the whole specimen, which now leads to a measurable but still small stiffness reduction : 0.5 % if total loss of transverse stiffness in the zero degree plies is presumed in the laminated plate theory and 0.25 % if a 50 % discount of transverse stiffness in these plies is presumed (9).

Along these longitudinal cracks failure of the $0°$-fibers is possible even before the actual fracture stress of the fibers is reached. The longitudinal cracks grow not only along one side of the fiber, but locally a bridging of fibers from one side of the crack to the other can occur. In Fig. 6 is shown such a situation, where a bundle of fibers is crossing a longitudinal crack.

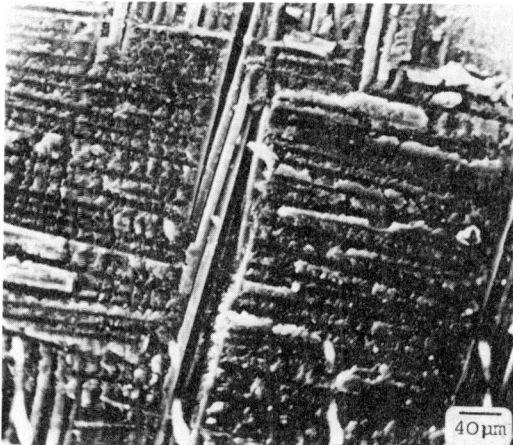

Fig. 6   SEM-micrograph. $0°$-fiber bridge crossing a longitudinal crack.

In this situation not only tensile stresses are acting into the longitudinal
direction (X-direction) of the fibers, but also as a result of the transverse
stresses (σ_2; compare Fig. 4) additional shear stresses are present.
This is schematically shown in Fig. 7. Because of their microstructure
carbon-fibers react very sensitive to shear stresses. This will be discussed
later in a separate chapter. It is now possible that failure of the 0°-fibers
occurs even before the actual failure stress of the fiber has been reached,
because shear stresses and tensile stresses act together on the fiber.
A superposition of these two stresses can now lead to fiber fracture at
loads which are lower than the expected tensile fracture stress.

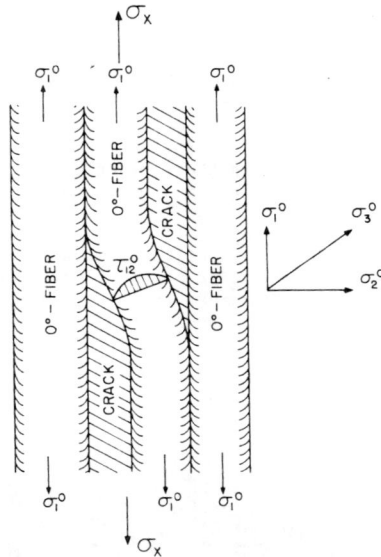

Fig. 7 Model of fiber fracture by in-plane cracks.

In Fig. 8a is shown a broken fiber at the edge of a longidutinal crack.
On one side is to be seen a small slit between a fiber and the crack surface which gives a hint that there a bridging of the crack has taken place.
Figure 8b shows a view perpendicular to the crack surface of a longitudinal
crack. The longitudinal crack was sectioned out of a fatigue specimen.
Clearly can be seen, especially at the top of Fig. 8b, that there is formed
a rather smooth fracture surface. This fracture surface has been developed
during the early stage of the fatigue life of a specimen. With increasing
crack propagation rate of the longidutinal crack the crack surface topography
becomes more irregular and looking more cleavage like. But the main topic
that can be recognized from Fig. 8b is, that a number of broken 0°-fibers
can be seen on the crack surface. The development of the 0°-fiber fracture,
as described above, will give an additional but secondary influence to the
reduction of the elastic module during fatigue in stage II.

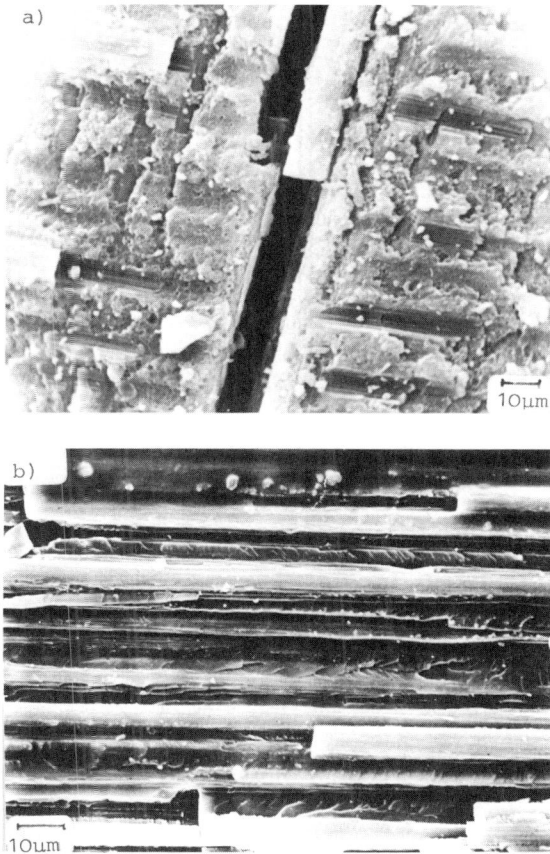

Fig. 8   SEM-micrograph.
a) Broken 0°-fiber at the edge of a longitudinal crack.
b) Broken 0°-fiber on the crack surface.

A further important damage mechanism, as a result of longitudinal cracks, can be observed in the interior of a cross ply laminate. By carefully investigating Fig. 3 it turns out that locally, at selected areas, delaminations developed in the interior at crossing-overs of longitudinal cracks with transverse cracks. This phenomenon of internal delamination has carefully been discussed in (9). The nature of their development is illustrated in Fig. 9. However, the local tensile stresses are provided by the longitudinal cracks. The tensile nature of the stress produces an out-of-plane tensile stress component, σ_t, at the interface.

90° 0°

LONGITUDINAL
CRACK

DELAMINATION

TRANSVERSE
CRACK

σ_t

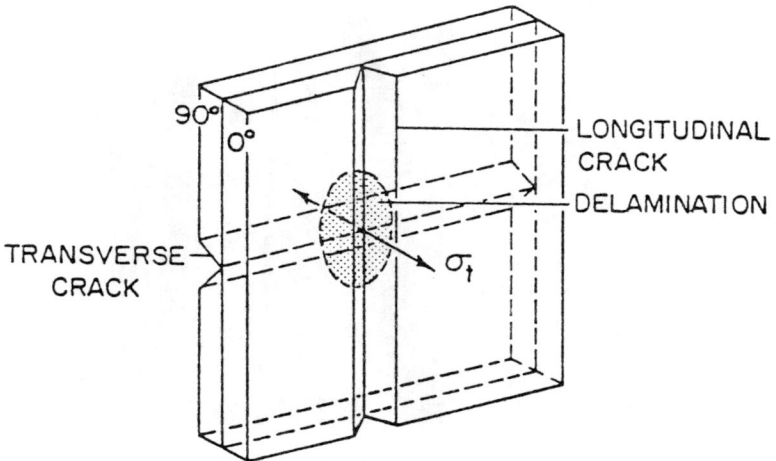

Fig. 9   Influence of longitudinal and transverse cracks on interior delamination formation (9).

In addition, an out-of-plane tensile stress component, σ_t, is provided by
the transverse cracks. At intersections of these cracks, the stresses are
additive and the propensity for the nucleation and growth of delaminations
should be highest. This is, in fact, observed experimentally. The centers
of incipient delaminations frequently coincide with crack intersections.
These are also the points of maximum delamination opening (9). However,
the damage mechanisms associated to longitudinal crack formation contribute
significantly to further stiffness reduction preparing the final life of a
specimen.
Figure 10 shows the stiffness reduction behaviour of a specimen which was
loaded at a level, where no development of longitudinal cracks could be
observed after more than 10^6 cycles (16). Following the initial decrease
in stiffness as a result of transverse crack formation no further variation
in stiffness was observed over the entire fatigue life.

   The damage mechanisms, which have been described up to now, lead to the
development of internal damage, and with that to a measurable reduction of
the stiffness of the different laminate types, but the mechanisms leading
to final failure of a laminate, that is the fracture of 0°-fibers (those
fibers which carry the main portion of the load), have not yet been described.
Up to now all those mechanisms are developed, which prepare the final failure of a specimen. Therefore a sharp separation of stage II and III is
not possible, the transition is more or less fluent. All the mechanisms
which are already developed during stage II, and which will lead in later
stages of the fatigue loading to increasing fiber fracture, will be discussed in the following.

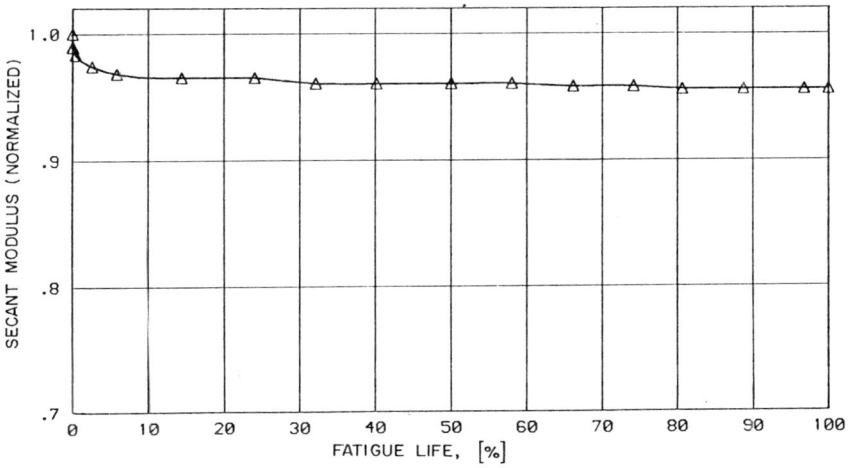

Fig. 10 Stiffness reduction curve for a $[0_2,90_2,0_2,90_2]_s$ -laminate (15).

Another important mechanism of fibre fracture could be recognized in cross ply laminates. The development of transverse cracks of the CDS leads in the neighbouring of 0°-fibers locally to an increase of the stresses. In the same time there are in the Z-direction (thickness direction) additional stresses (16). This leads to the situation that 0°-fibers near the tip of transverse cracks loose contact to their neighbouring 0°-fibers which then will lead to the formation of a bending moment acting on these fibers and, therefore, a superposition of shear and tensile stresses occurs, visible in fiber kinking. This again can lead to an earlier failure of the 0°-fibers (Fig. 11).

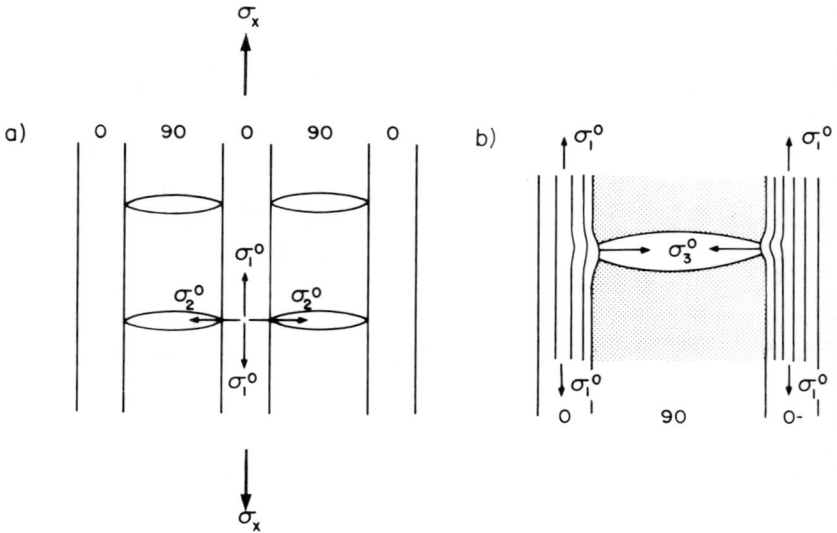

Fig. 11   a) Stress distribution in a [0,90,0]-laminate, if transverse cracks are present (schematically).
          b) Explanation of fiber fracture ; superposition of σ_1^0 and σ_3^0-stresses.

In Fig. 12a is shown in a scanning electron micrograph fiber fracture as the result of the transverse cracks of the CDS. The fiber fracture is concentrated at the straight line that crosses the whole width of the specimen, equal to the position of the crack tip of a transverse crack. In Fig. 12b is shown the same location with a lower magnification. The tips of broken 0°-fibers form equidistant lines (shown here by the darts) which have the same distance as the transverse cracks of the CDS. Figure 12c again shows broken 0°-fibers. Using the deply technique, it was possible to isolate fiber fracture prior to final failure of the specimen. In the case of Fig. 12c the specimen was loaded to about 80 % of its fatigue life. The then following deply technique made it possible to corroborate the results

shown in Fig. 12a and b. It was shown in (17) that the number of broken
0°-fibers starts early in a fatigue history and increases with increasing
fatigue life. But two fiber layers under the surface layer the number of
broken 0°-fibers is drastically reduced, what means, that the internal
cracks introduce fiber fracture, but their influence diminishes deeper in
a 0°-ply.

Fig. 12 SEM-micrographs. Fracture of 0°-fibers at the tip of transverse
cracks of the CDS.
a) High magnification.
b) Lower magnification; fiber fracture along equidistant lines
as they develop from the CDS.
c) Fiber breaks in 0°-fibers using the deply-technique.

If in a laminate ± 45°-plies are next to a 0°-ply, transverse cracks
under an angle of about 45° to the loading direction can be observed. Here
another failure mechanism for the fracture of the 0°-fibers can be observed.
As a result of the interlaminar shear stresses between the 0° and 45°-plies,
moving of the crack surfaces, as shown in Fig. 12, occurs. Because of the
shearing of the crack surfaces bending of the 0°-fibers prior to the crack
tip occurs. Also here a superposition of several stresses, as there are
the tensile stress in the loading direction and the shear stress in the

direction of the fibers in the 45°-ply, occurs.
A bending moment is acting on the 0°-fibers connected to the crack tip.
Therefore, also here 0°-fibers can fail before their fracture stress has been
reached. Figure 14a shows in a scanning electron micrograph the trace of
0°-fiber passing a transverse crack in a 45°-ply. It can be seen that the
trace of the 0°-fiber is shifted in that way that the upper part of the
trace has been moved a certain amount parallel to the lower part of the trace,
which led to the two breaks of the trace line. In Fig. 14b are shown several
broken fibers which are concentrated along the line of a transverse crack,
using the deply technique. Also this specimen was fatigued up to 80 % of its
total expected life.

It seems that failure of the 0°-fibers, at least at certain locations, as
described above, has some influence to the decrease of the stiffness during
fatigue life (18).

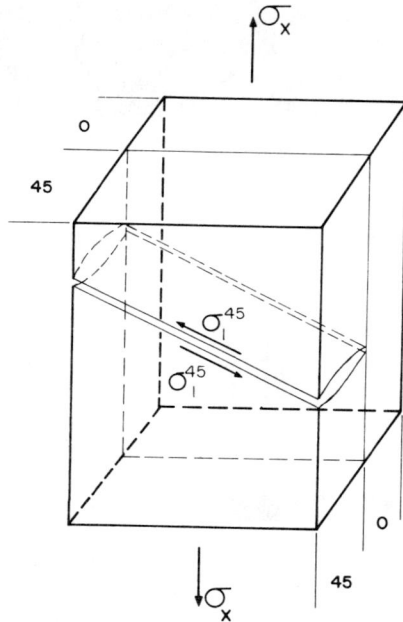

Fig. 13   Failure of the 0°-fibers in a [0,±45,0]-laminate as a result of
a transverse crack in the 45°-ply. Shearing of the crack surfaces
because of the tensile stresses (schematically).

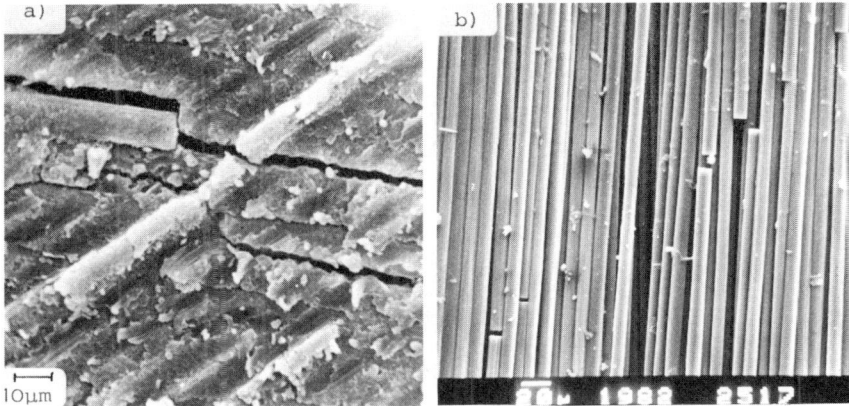

Fig. 14 SEM-micrograph. Fracture of 0° fibers at the tip of a micro-
crack. $[0,\pm45,0]_2$-laminate.
a) Trace of a 0°-fiber.
b) Fiber breaks on 0°-fibers using the deply-technique.

c) Stage III

In this section an explanation is tried for understanding the final failure of a composite laminate, which is announced by a sudden and rapid drop
in stiffness. Formation of longitudinal cracks and coalescence of delaminations between adjacent longitudinal cracks (compare Fig. 15), (9) lead
to the isolation of a strand of 0°-fibers from a remainder material. However,
this total separation occurs only frequently. More often strands of 0°-fibers,
formed by longitudinal cracks, remain to be in direct contact to the adjacent
90°-plies. Final rupture of the specimen can now be explained by the
following.

As shown above, failure of 0°-fibers can be determined in a composite laminate at those locations, where internal transverse cracks reach a 0°-ply
(compare Figs. 11-14). However, normally fiber fracture was observed in
singlets and doublets and only quite seldom in triplets (17). An example
for a triplet is shown in Fig. 16. If now a fiber is broken, load has to
be transferred into the neighbouring fibers with a certain load increase
in those fibers. If failure occurs in doublets, further load increase has
to be expected. Now the question arises, how many interacting fiber breaks
are necessary for initiation failure of a strand that was isolated from
the bulk by the internal cracks. It can be expected that the number of
fiber breaks next to each other has to increase to initiate failure. Batdorf
(19) has proposed that the number of local fiber breaks required to fail
a bundle is given by the ratio of Weibull modulus of the bundle to that of
the fiber. This suggests more than three fiber breaks to initiate failure
of the whole strand. This result coincides well with the results shown in
(9) and (17), where a "stair step" appearance in stage III of the characteristic stiffness reduction curve for the cross ply laminates was observed.

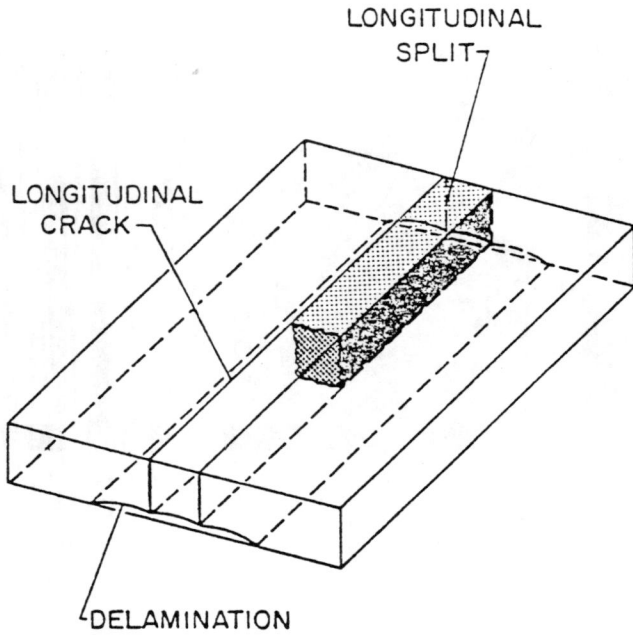

Fig. 15   Formation of strands of fibers by longitudinal cracks and internal delaminations (9).

Fig. 16   SEM-micrograph.   Fiber fracture, example of a triplet.

This means in summary, during fatigue in stage I and stage II a global damage pattern is produced being composed of transverse cracks, the nucleation and growth of longitudinal cracks, the production of internal and edge delamination and the initiation of fiber fracture at selected areas. In stage III now a transfer to local damage progression occurs, when the first multiplets in fiber fracture appeared leading to a strand failure which results in a sudden drop of the characteristic stiffness reduction curve and initiates final failure. This also is a reason for the large scatter in fatigue life and why the sudden drop in stiffness can hardly be predicted. The formation of multiplets occurs incidentally and not inevitably.

FIBER MICROSTRUCTURE AND FRACTURE

It was shown in the previous chapter that fiber fracture happened at those locations, where a local stress redistribution occurred, which alters the amount of load carried by each 0°-ply in that region. Such locations are in the vicinity of matrix cracks developed preferentially during fatigue loading along the fibers of zero degree or off-axis plies (in this investigation in ±45° and 90°-plies). In the vicinity of these cracks additionally to the tensile stresses also shear stresses act on 0°-fibers, because the micromechanical stress state is influenced by the formation of microcracks (compare Fig. 7, 11 and 13). A superposition of these stresses together with the fact that carbon fibers have pronounced anisotropic mechanical properties make fiber failure possible.

In Fig. 17 is shown the model of the fiber microstructure (after (20)). Carbon fibers have a lamellar arrangement of the carbon crystals with better oriented and larger crystallites in the skin, being essentially parallel to the surface, than in the core, where a complex interlinking and folding of the layer planes provides coherency over large cross-sectional areas of the fiber. The arrangement of the carbon crystals in the described way provides good mechanical properties in the fiber direction, but poor properties perpendicular to it (21).

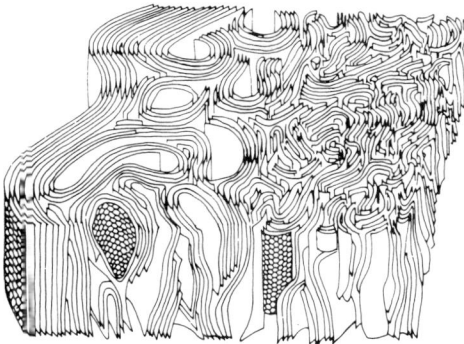

Fig. 17   Schematic model of the transverse structure in a carbon fiber (20).

If a carbon fiber is now embedded in a matrix, as it is the case in a composite, the fiber will be loaded mainly into the fiber direction, for the case of 0°-plies. In a composite without imperfections, loads perpendicular to the fiber will be homogeneously distributed over the entire fiber length. In the case of an imperfection (e.g. matrix cracks) load singularities occur in the vicinity of cracks and crack tips leading to a kinking of the fiber and in the region of kinking to the formation of shear stresses just into that direction where poor mechanical properties are present. These shear stresses superimpose the tensile stresses leading to fiber fracture. In Fig. 18a is shown an example of a broken fiber. Kinking of the fiber leads to fiber fracture in the area of high shear stresses (Fig. 18b). The poor mechanical properties perpendicular to the fiber direction can only be calculated from the fiber microstructure (20,21) and the lattice structure of a carbon crystal (2) rather than from test results. Only one publication is known to the author where some mechanical properties transverse to the fiber direction were investigated experimentally (22) and where the anisotropic mechanical properties were corroborated.

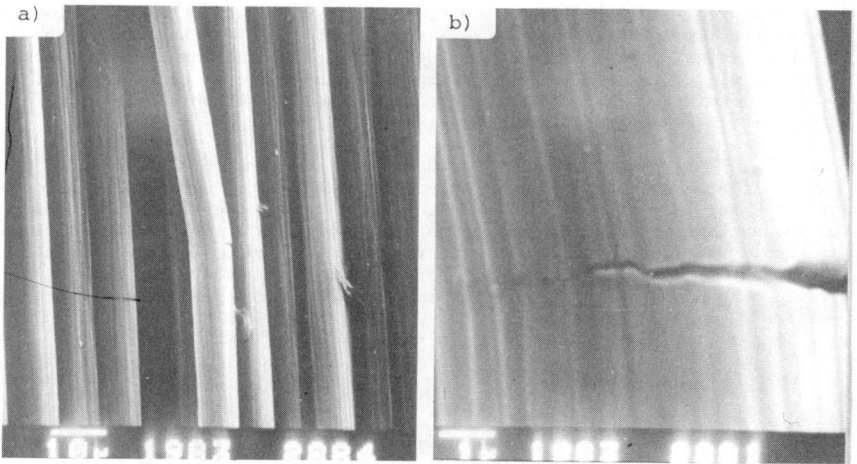

Fig. 18   SEM-micrographs.
a) Bending of the fiber back to fiber fracture in the area of high shear stresses.
b) High magnification of a).

REFERENCES

1. Zweben, C., Composites 12, pp. 235-240, 1981.

2. Böder, H., Gölden, D., Rose, Ph. and Würmscher, H., Z. Werkstoff-
    technik, 11, pp. 275-281, 1980 (in German).

3. Schulte, K., Henneke, E.G., Duke, J.C., Lecture Preprint DGM-Conf. :
    Verbundwerkstoff-Stoffverbunde, 9-11 May 1984, to be published in
    in Z. Werkstofftechnik 1985 (in German).

4. Reifsnider, K.L., Schulte, K. and Duke, J.C., ASTM-STP 813, pp. 136-
    159, 1983.

5. Reifsnider, K.L. and Jamison, R.D. : Int. J. Fatigue, pp. 187-197,
    1982.

6. Schulte, K., Reifsnider, K.L. and Stinchcomb, W.W., Proceeding 18th
    AVK-Conference, Oct. 5-7, 1982, Freudenstadt, W.-Germany, pp. 29-1 -
    29-8 (in German).

7. Reifsnider, K.L., 14th Annual Society of Eng. Science Meeting, Nov.
    14-16 (1977), Lehigh University, Bethlehem, PA, U.S.A.

8. Reifsnider, K.L. and Highsmith, A.L., Materials : Experimentation
    and Design in Fatigue, Westbury House, Guildford, England, pp. 246-260,
    1981.

9. Jamison, R.D., Schulte, K., Reifsnider, K.L. and Stinchcomb, W.W.,
    ASTM-STP 836, pp. 21-55, 1984.

10. Bailey, J.E., Curtis, P.T. and Parvizi, A., Proc. R. Soc. Cond. 366,
    pp. 599-623, 1979.

11. Highsmith, A.L. and Reifsnider, K.L., ASTM-STP 797, pp. 103-117,
    1982.

12. Wang, A.S.D., Composite Technology Review 6, pp. 62-84, 1984.

13. Fukunaga, H., Chou, T.-W., Peters, P.W.M. and Schulte, K., Journal
    of Composite Materials 18, pp. 339-356, 1984.

14. O'Brien, T.K., ASTM-STP 775, pp. 140-167, 1982.

15. Ruttert, D., Studienarbeit Universität - GH Essen, 1984 (in German).

16. Reifsnider, K.L. and Talug, A., Int. Journal of Fatigue 1, pp. 3-11,
    1980.

17. Jamison, R.D., Ph.-D. Thesis, Virginia Polytechnic Institute and
    State University, Blacksburg, VA, U.S.A., 1982.

18. Steif, P.S., Stiffness Reduction Due to Fiber Breakage, Journal of
    Composite Materials 17, pp. 153-172, 1984.

19. Batdorf, S.B., J. Reinforced Plastics a. Composites 1, pp. 153-164, 1982.

20. Benett, S.C. and Johnson, D.J., Proceeding 5th Conference on Carbon and Graphite, pp. 377-386, 1978.

21. Fitzer, E. and Jäger, H., Lecture Preprint for Presentation of R. Weisz, F.A. Verbundwerkstoffe, München, 1984 (in German).

22. Smith, R.E., J. Appl. Phys. 43, pp. 2555-2561, 1972.

ADHESION PHENOMENA BETWEEN PHASES IN COMPOSITES : THE UNFOLDING MODEL

P.S. THEOCARIS

National Technical University of Athens, Greece

SUMMARY

In this paper a theoretical model was introduced for the evaluation of the mesophase layer based on the requirement of a smooth transition on the mechanical properties between phases of the composite. According to this model it was assumed that the physical properties of the mesophase unfold from those of the softer matrix. Thus, a multilayer model was assumed, improving the classical two-layer model introduced by Hashin and Rosen for the representative volume element of the composite.

The extent of the mesophase layer was defined by imposing to the model to satisfy the boundary conditions at the outer layers of the pseudophase with variable mechanical properties so that its properties match those of the external layers of the inclusions and the internal layers of the matrix. By measuring the elastic modulus of the composite and matching it with its value derived from the model the law of variation of the mesophase modulus may be accurately defined.

The model applies equally well for fibrous composites and particulates. However, in this paper the case of unidirectional fiber composites was treated and experimental results were presented.

INTRODUCTION

The adhesion between matrix and inclusions (fibers or particulates) in a composite material is among the main factors characterizing the mechanical and physical behaviour of the composite materials. All theoretical models describing these substances neglect to consider the influence of the boundary layer developed between phases during the preparation of the composite (1).

Around an inclusion embedded in a polymeric matrix a rather complex situation develops, consisting of areas of imperfect bonding, permanent stresses due to shrinkage, high stress-gradients or even stress-singularities, due to the geometry of the inclusions, voids, microcracks etc. Moreover, the interaction of the surface of the filler with the matrix is usually a procedure much more complicated than a simple mechanical effect. The presence of a filler actually restricts• the segmental and molecular mobility of the polymeric matrix, as adsorption-interaction in polymer surface-layers into filler-particles occurs. Then, the polymeric matrix, cast around the inclusions, creates phenomena of physical and chemical adsorption.

Physisorbed (physically adsorbed) layers of the matrix contribute, in general, to weak mesophases. However, the physical interpenetration of the boundary layer of the matrix in cavities and other rough regions of the surfaces of solid inclusions, interrelated with the restrained development of the molecular structure of the polymeric chains there, and any other structural variations of the adjacent layers, create an intermixing and interpenetrating phenomenon, which influences considerably the molecular structure of the mesophase, thus resulting to variations of its mechanical strength. Thus, the mechanical properties of the matrix films and layers close to the interfaces, are strongly depending on the physical situation of this boundary layer.

On the other hand, chemisorbed (chemically adsorbed) molecules on the interfaces create structural variations, by developing beaded structures of caged molecules or ladder-like molecules. All these types of chemisorbed elements on inclusions lead to rapid variations of the properties and mechanical strength of the interface layer, close to the surface of inclusions.

The above sketched phenomena happening around an inclusion create an intermediate boundary layer of variable thickness. However, this zone is extended beyond the thin layer including the phenomena of physisorption and chemisorption, and it incorporates the zones of imperfect bonding and shrinkage stresses, the high stress gradients, or even stress singularities, due to geometric discontinuities of the surfaces of inclusions, to the concentration of voids and to the impurities, microcracks and other anomalies.

In this study the existence of the boundary layer, constituting the mesophase, and developed between the two main phases of a two-phase composite was taken into account for the development of a convenient model describing the mechanical behaviour of composites and especially unidirectional fiber composites and laminates. This layer was assumed as developed entirely on the side of the softer polymeric matrix, and the harder inclusion is considered as neutral. Moreover, the mesophase was assumed as an independent pseudo-phase of variable properties, matching those of the inclusion on the one side, and the matrix on the other. The model was founded on the same basic ideas as the Hashin-Rosen model (2).

THE UNFOLDING MODEL

The evaluation of the characteristic properties of the mesophase was achieved by introducing an improved law of mixtures between phases incorporating the influence of the third phase. It was further assumed that the mechanical and the physical properties of the mesophase "unfold" from those of the hard-core filler to those of the softer matrix.

In the following and for reasons of brevity we shall concentrate to the case of unidirectional fibrous composites, giving all the relevant theory for these substances. However, the same theory holds also for particulates with relations much more complicated and referring to compliances instead of moduli (1).

For a three-layer model of a fibrous composite where the mesophase occupies the intermediate phase, the modulus of composite, E_c, may be expressed by an improved law of mixtures as follows :

$$E_c = E_f v_f + E_i^a v_i + E_m v_m \qquad (1)$$

The indices f, m and i denote filler, mesophase and matrix respectively, and for the three-volume fractions it is valid that :

$$v_f = (\frac{r_f}{r_m})^2$$

$$v_m = (\frac{r_i^2 - r_f^2}{r_m^2}) \tag{2}$$

$$v_i = (\frac{r_m^2 - r_i^2}{r_m^2})$$

with

$$v_f + v_m + v_i = 1 \tag{3}$$

Furthermore the E_i^a value for the variable elastic modulus of the mesophase corresponds to its mean-value. While this law of mixtures is adequate for uniaxial fibrous composites its validity to particulates is doubtful. For a discussion of this problem see ref. (1).

In relation (1) the modulus of mesophase, E_i, was initially defined by the two-term unfolding model (1), expressed by :

$$E_i(r) = E_f(\frac{r_f}{r})^{2\eta} + [E_m - E_f(\frac{r_f}{r_i})^{2\eta}] \frac{r-r_f}{r_i-r_f} \tag{4}$$

However, it is more efficient and simplifies considerably the amount of calculations if we use instead of relation (4) a recent modification given by (5) :

$$E_i(r) = E(\frac{r_f}{r})^{\eta^*} (\frac{r_i-r}{r_i-r_f}) + E_m [\frac{r_i(r-r_f)}{r(r_i-r_f)}] \tag{5}$$

In both last relations, the polar distance, r, varies between r_f and $r_i = r_f + \Delta r$, where r_i is the external radius of the mesophase, whereas the exponents η and η^* define the law of variation of this modulus. The improved model defined by Eq. (5) yields for a typical $E_i = f(r)$ variation of the modulus of the mesophase, the curve indicated in Fig. 1, where an abrupt change in its slope appears at the vicinity of $r = r_f$. In reality, the material, which constitutes the mesophase, is almost the same as the polymeric matrix, but with the differences stated in the introduction. Thus, it is logical to expect that the larger part of the thickness of the mesophase has an elastic modulus closer to the value E_m of the matrix modulus. Moreover, it is logical not to expect any slope difference in $E_i(r)$ at the vicinity of $r = r_i$ and to expect a smooth transition between the $E_i(r)$-variable modulus and the E_m-modulus. This implies that at $r = r_i$ the tangent to the $E_i = f(r)$ and $E_m = \varphi(r)$ curves should be common.

This yields :

$$E_i^a = \frac{E_m}{\Delta r_i^2} [\ \Delta r_i^2 - r_f r_i\ (\ell n\ r_i - \ell n\ r_f - 1) + r_f^2\ ]$$

$$+ \frac{E_f}{\Delta r_i^2} \left| \frac{r_i^2 (\frac{r_f}{r_i})^{n^*} - r_i r_f}{(1 - n^*)} + \frac{r_f^2 - (\frac{r_f}{r_i})^{n^*} r_i^2}{2 - n^*} \right| \tag{8}$$

where $\Delta r_i = (r_i - r_f)$.

Introducing the value for E_i^a given by Eq. (8) into relation (1) and measuring the value of the modulus E_c of the composite it is possible, by solving numerically the system of Eqs. (1) and (7), to evaluate the quantities n^* and r_i.

RESULTS AND DISCUSSION

In order to check the consistency and the merits of the proposed analytical method for the evaluation of r_i, with results of the experimental method proposed by Lipatov, use was made of experimental data contained in ref. (1).

For the case of fibrous composites, the experiments were made with E-glass fiber-epoxy resin composites, with E-glass fibers having a radius of $r_f = 6.10^{-6}$ m and for filler volume fractions v_f varying between 10 and 70 percent. The elastic moduli of the fibers and the epoxy-resin were respectively : $E_f = 71,94 . 10^9$ Pa and $E_m = 3,445 . 10^9$ Pa. Table I yields the experimental data for r_i and E_c taken from ref. (1). Using these values for r_f, v_f and E_c as input data in the method described in this paper we have solved numerically the set of equations (11) and (16). From the solution of this set of equations we calculated the respective values for r_i, which are also shown in Table I.

The improved two-term unfolding model may be equally well applied for particulates as it has been applied in this paper to fibrous composites. Indeed this model was used in ref. (3) for evaluating the thickness of mesophase in particulates. In this case, the two models, the simple and the improved one, yielded almost the same values for r_i. However, when these models are compared with the experimental evidence by using calorimetric measurements and Lipatov's theory (4) for iron-epoxy particulates it was found that the percentage error in the prediction of r_i is greater for low volume fractions of the filler. This result is due to the fact that volume fractions of the mesophase are low too in these cases, and the equation expressing the compliance of the composite in terms of the compliances of the constituent phases is not sensitive to small changes of r_i.

Finally, it may be derived from these results that the results yielded by the proposed two-term unfolding model are in satisfactory agreement with the experimental values for r_i derived from the theory of Lipatov (4).

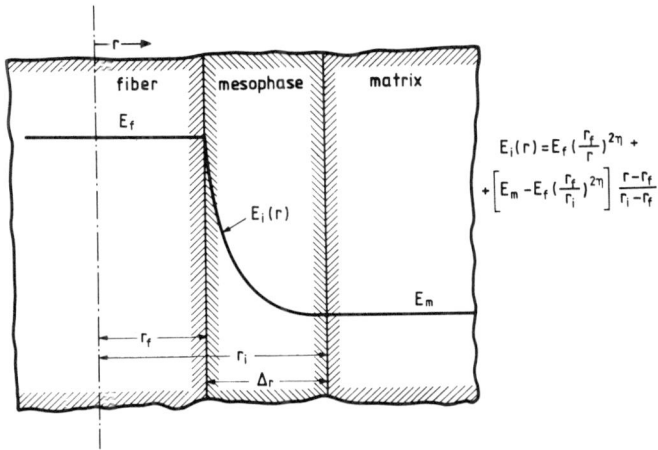

$$E_i(r) = E_f \left(\frac{r_f}{r}\right)^{2\eta} + \left[E_m - E_f\left(\frac{r_f}{r_i}\right)^{2\eta}\right] \frac{r - r_f}{r_i - r_f}$$

Fig. 1 Variation of the $E_i(r)$-modulus with respect to the thickness of the mesophase, accordingly to the initial form of the 2-term unfolding model.

Then, it is valid that :

$$\left.\frac{dE_i(r)}{dr}\right|_{r=r_i} = 0 \tag{6}$$

Differentiating relation (5), putting $r = r_i$ and introducing the result into Eq. (6) the following relation may be derived :

$$\eta^* = 1 + \frac{\ln(E_m/E_f)}{\ln(r_f/r_i)} \tag{7}$$

Relation (7) which interconnects, in a simple manner, the ratio of the logarithms of the ratios of the moduli of the constituent phases and the radii of inclusions and the mesophase of the composite constitutes the main advantage of the improved version of the two-term unfolding model.

For this improved two-term model, the average elastic modulus, E_i^a, for the mesophase material may be found by integrating relation (5) from $r = r_f$ to $r = r_i$ and taking the average value of this definite integral over the same range.

Moreover, in Fig. 2 the values of the mesophase thicknesses derived from the two-term simple and improved models are plotted, and as it is seen, the two theoretical predictions formed a narrow band, as lower and upper bounds, which contains almost all the experimental results.

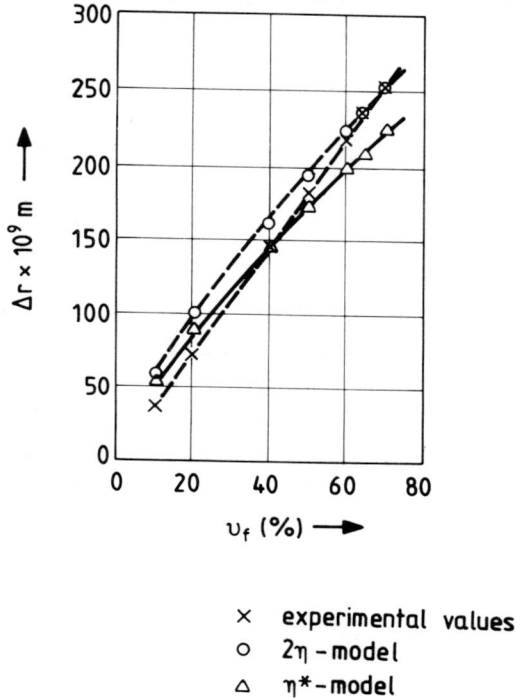

× **experimental values**
○ **2η – model**
△ **η* – model**

Fig. 2   Experimental data for the mesophase radius r_i and theoretical predictions from the initial and modified 2-term unfolding models for the case of fibrous composites.

CONCLUSIONS

A simple and effective model for evaluating the thickness of the mesophase in fibrous and particulate polymeric composites was established in this paper.   The results are more encouraging for the case of fibrous composites, as a result of the simplicity of the law of mixtures applying in their case. For particulates the results are also in less satisfactory agreement with the experimental values for the thickness of the mesophase.

The model introduced in this paper is based on principles of physics and mechanics, and it combines, in a harmonious way, the properties of the main phases together with the experimentally defined properties of the composite. Moreover, the model presents a high degree of flexibility and may be adapted easily to real situations.

Table 1 Experimentally obtained values for r_i, ΔC_p and E_c of E-glass fiber-epoxy resin composites with E-glass fibers having a radius of $r_f = 6 . 10^{-6}$ m.

υ_f %	r_i ×10^6 m	E_c ×10^-9 Pa	E_i^a ×10^-9 Pa	r_i ×10^6 m	η^*	$\dfrac{r_i(th)-r_i(ex)}{r_i(ex)}$ ×100
	DSC-Tests [1,4]		η^* - two - term model			
10	6.036	13.32	17.92	6.0527	348.503	0.276
20	6.071	17.23	17.89	6.088	209.714	0.28
40	6.145	31.12	17.839	6.143	130.02	0.032
50	6.182	38.11	17.81	6.172	108.52	0.161
60	6.217	45.12	17.78	6.198	94.59	0.305
65	6.235	48.62	17.77	6.207	90.59	0.449
70	6.254	52.15	17.76	6.223	84.27	0.495

Finally, it is worthwhile also noting that the proposed theoretical method can be used in the inverse way, that is for the prediction of the elastic modulus of the composite, E_c, by using the experimental values for the mesophase radius, r_i detected by various experimental methods.

REFERENCES

1. Theocaris, P.S., The interphase and its influence on the mechanical properties of composites, New developments in the characterization of polymers in the solid state. Advances in Polymer Science, H.H. Kausch, ed., Springer Verlag, vol. 66, p. 200, 1984.

2. Hashin, Z. and Rosen, B.W., Journal of Applied Mechanics 31(3), 223, 1964.

3. Theocaris, P.S., Journal of Reinforced Plastics and Composites, vol. 3, 3, p. 204, 1984.

4. Lipatov, Yu. S., Physical chemistry of filled polymers. Translated from the Russian by R.J. Moseley, International Polymer Science and Technology Monograph no. 2, Originally published "Khimiya", Moscow, 1977.

5. Theocaris, P.S. and Phillipidis, Th., Journal of Reinforced Plastics and Composites, vol. 4, no. 4, p. 200, 1985.

B. CONTRIBUTIONS

THE MECHANICAL BEHAVIOUR OF ELASTIC-PLASTIC FIBER-REINFORCED LAMINATED PLATES

J. ABOUDI and Y. BENVENISTE

Tel-Aviv University, Ramat-Aviv, Tel-Aviv, Israel

INTRODUCTION

The analysis of elastic fiber reinforced laminated plates is by now well established. Constitutive laws representing the fiber reinforced medium are available and these have been incorporated in several lower and higher order plate theories (see Christensen [1]). The number of investigations on the mechanical behaviour of inealstic fiber reinforced laminated on the other hand are rather scarce. Here, each lamina is made of elasto-plastic matrix and fiber constituents and there is a lack of effective constitutive laws describing such unidirectional media. Furthermore even if the behaviour of each lamina is determined, due to the inelastic behaviour there remains essential difficulties in determining the stress field distribution across the laminated plate.

In elastic plate theories once the strain field is prescribed in the framework of an appropriate plate theory, the stress field immediately results by the explicit strain-stress relations of the lamina. In inelastic laminates however, such a direct determination of the stress field from the strain distribution is not possible due to the complexity of the non-linear constitutive relations. Furthermore although the Love-Kirchhoff hypothesis may be adequate in describing the deformation field across the plate, the stress distribution is necessarily nonlinear and needs to be determined by an appropriate method.

The present work is concerned with the determination of the constitutive laws describing the gross behaviour of inelastic fiber reinforced laminated plates. The fiber reinforced lamina is modeled according to the theory presented in [2]. In this work the average behaviour of unidirectional fiber reinforced composites whose constituents are elasto-plastic have been formulated. The matrix is elasto-plastic work-hardening and the fiber material is represented by an anisotropic constituent because of the widespread use of carbon and graphite fibers which are highly anisotropic. Comparisons with other ananlytical and experimental approaches has proven in [2] the reliability of the obtained constitutive laws.

The determination of the gross behaviour of the laminated plate in bending and stretching deformations is carried out in the framework of the Love-Kirchhoff hypothesis for the strain field. The stresses on the other hand are expressed as a series of Legendre polynomials in a local coordinate system defined in each lamina [3]. The net axial forces and moments can readily be obtained in terms of the coefficients of the stress series. Using an averaging procedure on each lamina and employing the orthogonality of the Legendre polynomials it is possible to obtain relations between the coefficients in the stress series and the stretchings and curvatures at the reference surface of the composite plate. A system of equations is thus derived which describes the variation of the stress field across the laminated fiber reinforced plate, once the Love-Kirchhoff assumption is adopted. These equations furnish at the same time the gross constitutive behaviour of the plate, that is, provide relations between the resultant forces and moments and the reference plane stretchings and curvatures. Due to the presence of inelastic effects, the implementation of these relations can only be carried out incrementally in time. In the process of solving boundary value problems the obtained constitutive relations have of course to be supplemented by the equilibrium or dynamical equations of the plate.

THE EFFECTIVE CONSTITUTIVE BEHAVIOUR OF THE LAMINATED PLATE

A laminated plate is considered in which every lamina has a width of h_k with $k = 1,2,\ldots,K$, where K represents the total number of layering. For combined bending and stretching deformations the displacements are described in the framework of the Love-Kirchhoff theory

$$u_x(x,y,z,t) = u_o(x,y,t) - z\,\frac{\partial w_o}{\partial x}\,(x,y,t)$$

$$u_y(x,y,z,t) = v_o(x,y,t) - z\,\frac{\partial w_o}{\partial y}\,(x,y,t) \tag{1}$$

$$u_z(x,y,z,t) = w_o(x,y,t)$$

where the functions u_o, v_o and w_o are the displacements at a reference surface $z = 0$ of a coordinate system (x,y,z) with z perpendicular to the layering and t denotes the time.

A local coordinate system is located at the midplane of every lamina and defined as $(x,y,z^{(k)})$ with $k = 1,\ldots,K$. Introducing the nondimensionalization,

$$\zeta^{(k)} = z^{(k)} / (h_k/2) \tag{2}$$

and employing the Love-Kirchhoff expressions for the displacements, the following form for the strains is obtained :

$$\underset{\sim}{\varepsilon}^{(k)} = \underset{\sim}{e}^{(k)}_{(o)} + \sqrt{3}\,\underset{\sim}{e}^{(k)}_{(1)}\,\zeta^{(k)} \tag{3}$$

where

$$\underset{\sim}{\varepsilon}^{(k)} = \{\varepsilon^{(k)}_{xx(o)},\ \varepsilon^{(k)}_{yy},\ \varepsilon^{(k)}_{xy}\}$$

$$\underset{\sim}{e}^{(k)}_{(o)} = \{\varepsilon^{(k)}_{xx(o)},\ \varepsilon^{(k)}_{yy(o)},\ \varepsilon^{(k)}_{xy(o)}\} \tag{4}$$

$$\underset{\sim}{e}^{(k)}_{(1)} = \{\varepsilon^{(k)}_{xx(1)},\ \varepsilon^{(k)}_{yy(1)},\ \varepsilon^{(k)}_{xy(1)}\}[\,h_k/(2\sqrt{3})]$$

and no summation is implied on k.

Every layer of the laminate is now modeled according to the effective constitutive law developed by Aboudi [2] for inelastic fiber reinforced composite
materials. In that work the matrix is isotropic in the elastic range and iso-
tropically work-hardening in the plastic range. The fibers on the other hand
are elastic but exhibit pronounced transverse isotropic behaviour.
The metal matrix is represented by the unified elastic-plastic theory of
Bodner and Parton [4] which has the convenient property that no separate spe-
cification of a yield criterion is required nor is it necessary to consider
loading and unloading separately.

According to the constitutive law of [2], and for the special case of plane
stress, the average stresses and strains are related by

$$\underset{\sim}{\varepsilon}^{(k)} = \underset{\sim}{E}^{(k)} \ \underset{\sim}{\sigma}^{(k)} + \underset{\sim}{F}^{(k)} \tag{5}$$

where $\underset{\sim}{\sigma}^{(k)}$ is defined by

$$\underset{\sim}{\sigma}^{(k)} = \{\sigma_{xx}^{(k)}, \ \sigma_{yy}^{(k)}, \ \sigma_{xy}^{(k)}\} \tag{6}$$

$\underset{\sim}{E}^{(k)}$ is a 3x3 matrix which consists of the effective elastic (initial) moduli
of the composite and $\underset{\sim}{F}^{(k)}$, denoted by :

$$\underset{\sim}{F}^{(k)} = \{F_{xx}^{(k)}, \ F_{yy}^{(k)}, \ F_{xy}^{(k)}\} \tag{7}$$

represents the inelastic strains. A detailed description of $\underset{\sim}{E}^{(k)}$ and $\underset{\sim}{F}^{(k)}$
can be found in [3] in which the fiber material was taken to be isotropic.
The generalization to the anisotropic case is obtained by using the formulation of [2].

As discussed in the introduction, in an inelastic laminated plate the
distribution of the stresses across the thickness is certainly nonlinear in
stretching and bending deformations, and needs to be determined by an appropriate method. A legendre expansion formalism is used in this work to develop the gross constitutive relations of the laminated plate. Let us assume
the following expansion for the stresses :

$$\underset{\sim}{\sigma}^{(k)} = (1+2n)^{1/2} \ \underset{\sim}{\tau}_{(n)}^{(k)} \ P_n(\zeta^{(k)}) \tag{8}$$

where $\underset{\sim}{\tau}_{(n)}^{(k)}$ defined by

$$\underset{\sim}{\tau}_{(n)}^{(k)} = \{\tau_{xx(n)}^{(k)}, \ \tau_{yy(n)}^{(k)}, \ \tau_{xy(n)}^{(k)}\} \tag{9}$$

are the coefficients of the expansion in the Legendre polynomials $P_n(\zeta^{(k)})$.
In the above equation summation is implied by the repeated index $n = 0,1,2,...N$
and there is no sum on k. Equation (8), together with (3), is now introduced
in (5), each side multiplied by $P_m(\zeta^{(k)})$, and then integrated between -1
and $+1$ with respect to $\zeta^{(k)}$. Making use of the orthogonality property of
the Legendre polynomials yields :

$$e_{\sim(m)}^{(k)} = E_{\sim}^{(k)} \; \tau_{\sim(m)}^{(k)} + R_{\sim(m)}^{(k)} \tag{10}$$

where

$$R_{\sim(m)}^{(k)} = [(2m+1)^{1/2} / 2] \int_{-1}^{+1} F_{\sim(m)}^{(k)} \; P_m(\zeta^{(k)}) \; d\zeta^{(k)} \tag{11}$$

in which $e_{\sim(m)}^{(k)} = 0$ for $m \geqslant 2$.

Again no sum is implied on k and m. Solving equation (10) for $\tau_{\sim(m)}^{(k)}$ yields :

$$\tau_{\sim(m)}^{(k)} = C_{\sim}^{(k)} \; (e_{\sim(m)}^{(k)} - R_{\sim(m)}^{(k)}) \tag{12}$$

where $C_{\sim}^{(k)} = (E_{\sim}^{(k)})^{-1}$. This equation determines the coefficients $\tau_{\sim(m)}^{(k)}$ once the coefficients $e_{\sim(m)}^{(k)}$ and the functions $R_{\sim(m)}^{(k)}$ are known at any instant. The field variables of the laminated plate are determined by an incremental procedure which is explained in [3].

Having thus obtained constitutive relations between the coefficients in the stress series (8) and the coefficients in the expressions for the strains (3), the gross constitutive relations of the plate relating the net moments and forces to the reference plane stretchings and curvatures can readily be obtained.

RESULTS

In the present work the developed theory is illustrated by the simplest possible loading of the plate, that is, pure cylindrical bending. Each lamina is made of aluminium matrix (2024-T4) and reinforced by graphite fibers (T-50).
The behaviour of the aluminium alloy in a simple tension test at room temperature is shown in Fig. 1 (Young's modulus 72.4 GPa and Poisson's ratio $\nu = 0.33$). The specific inelastic parameters are given in [2]. The T-50 graphite elastic fibers are transversely isotropic (axial Young's modulus 388.2 GPa, axial Poisson's ratio 0.41, transverse Young's modulus 7.6 GPa, transverse Poisson's ratio 0.45, axial shear modulus 14.9 Gpa). In Fig. 2, the moment curvature relations of three-layered cross-ply plates in pure cylindrical bending is shown. M^* is the non-dimensional moment per unit length defined by $M^* = M/[(E)_{A1} h^2/2]$, where h is the thickness of each lamina and κ^* is the non-dimensional curvature defined by $\kappa^* = \kappa h/2$.

The above results are an illustration of the presented theory for the case of pure cylindrical bending. It is clear that, having obtained the gross constitutive behaviour of the laminate, the theory can be applied to more general problems of inelastic laminated plates involving stretching and bending deformations.

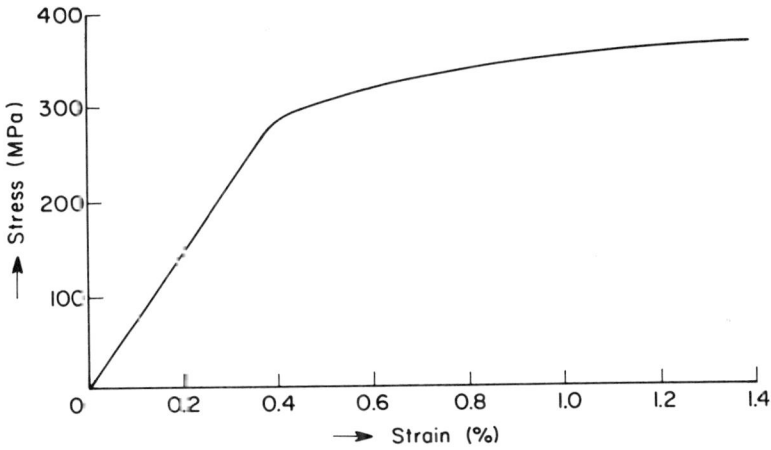

Fig. 1 The behaviour of aluminium alloy in a simple tension test.

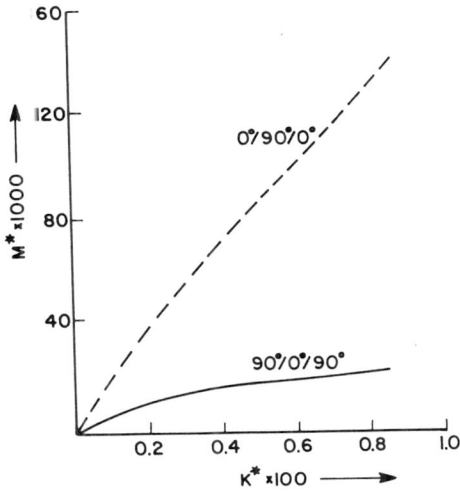

Fig. 2 Moment-curvature relations of three-layered cross-ply plates in pure
cylindrical bending.

REFERENCES

1. Christensen, R.M., Mechanics of Composite Materials, Wiley, New York, 1979.

2. Aboudi, J., Effective Behaviour of Inelastic Fiber Reinforced Composites, Int. J. Engng. Sci. vol. 22, no. 4, pp. 439-449, 1984.

3. Aboudi, J. and Benveniste, Y., Constitutive Relations for Fiber Reinforced Plates, J. of Appl. Mech., vol. 51, pp. 107-113, 1984.

4. Bodner, S.R. and Partom, Y., Constitutive equations for elastic-visco-plastic strain hardening material. J. Appl. Mech. 42, pp. 385-389, 1975.

THE APPLICATION OF KINETIC FRACTURE MECHANICS TO LIFE PREDICTION FOR
POLYMERIC MATERIALS

R.M. CHRISTENSEN and R.E. GLASER

Lawrence Livermore National Laboratory, California, U.S.A.

ABSTRACT

A recently developed theory of kinetic crack growth is applied to the
problem of the time-dependent failure of polymeric materials subject to
a timewise constance state of stress. The theory is generalized to include
a rather complete treatment of statistical variability. When restricted
to the power law range of behaviour, it is found that an assumption of a
Weibull distribution of static strengths implies rigorously that the corresponding lifetime distributions are also Weibully distributed. In fact,
an extremely simple interrelation is found whereby the shape parameter for
lifetime is equal to the product of the shape parameter for static strength
times the magnitude of the log stress versus log lifetime slope of the scale
parameter for the lifetime distribution. These theoretical predictions are
compared with the experimental results available from the testing of a particular polymeric fiber composite material, Kevlar 49/epoxy.

The lifetime stress rupture experimental data are shown by the solid line
segments in Fig. 1. The data are arranged in the following quantiles of failure : 2 %, 5 %, 8 %, 50 % and 90 %. Straight line segments are used to connect
the statistical failure quantiles at the sevel stress levels. The form
in Fig. 1 involves a linear scale for stress level versus a $\log_{10}$ time
scale for lifetime to failure. If the scales were both of logarithmic nature, a power law form would give a straight line representation. In fact,
the data in the upper portion of Fig. 1 exhibit a power law behaviour.
However, there is a strong downturn in the data, which is suggestive of a
separate chemical degradation mechanism. The theoretical predictions are
shown as the dashed curves in Fig. 1. These predictions are obtained in
the following manner. The experimental data at levels above 80 % of static
strength (intrinsic/static strength scale parameter) show a predominantly
power law behaviour. Accordingly, the corresponding parameters in (18)
and (19) are estimated to fit these upper level data. The method of maximum likelihood is used.

The main purpose of any theory is to make predictions outside the range
of accessible data. Figure 2 shows the use of the present theory to extend
the data base to examine extreme reliability levels. Failure quantiles
of 10^{-6}, 10^{-3} and 0.5 reveal the associated lifetimes at these reliability
specifications, as a function of stress level. For example, at about 50 %
of static strength load level, the 10^{-6} quantile gives an estimated lifetime of a little more than a year. It is seen that at the longer times
the chemical degradation mechanism begins to overtake the purely mechanical
effect of damage growth and the spread between the various reliability
levels diminishes relative to the log scale.

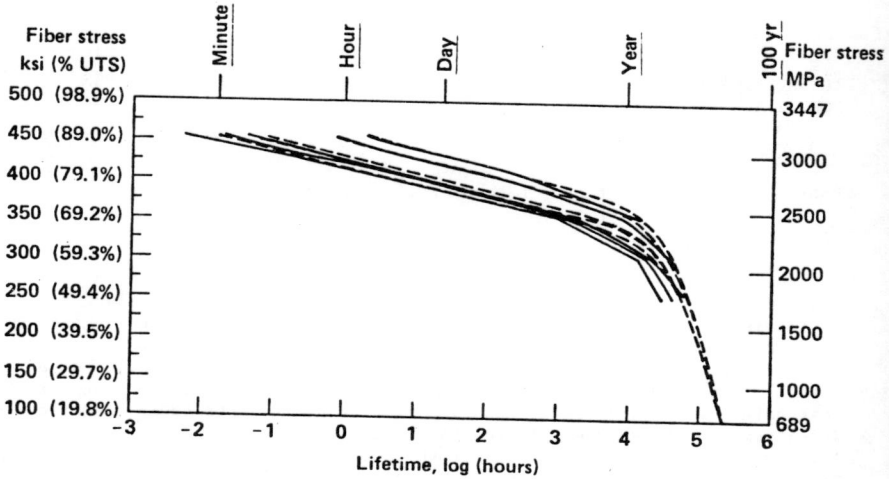

Fig. 1   Annual chemical degradation rate : 5 %
        Quantiles : 2.e-02, 5.e-02, 8.e-02, 5.e-01, 9.e-01.

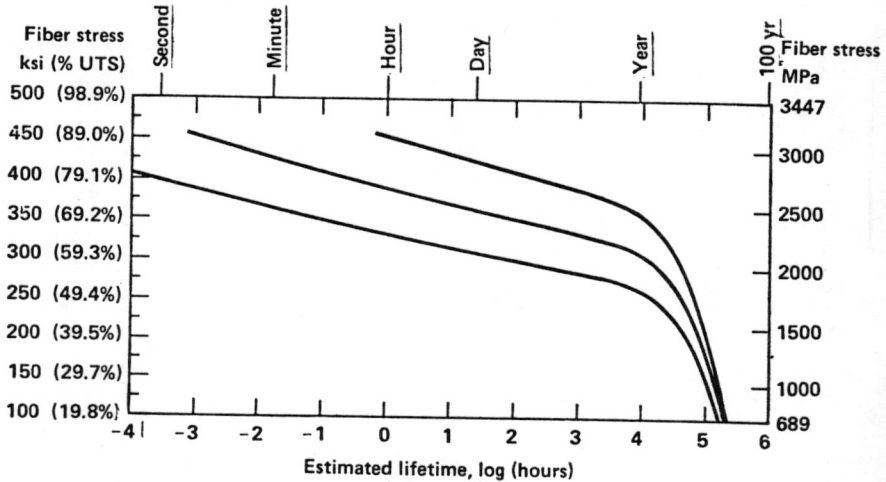

Fig. 2   Annual chemical degradation rate : 5 %
        Quantiles : 1.e-06, 1.e-03, 5.e-01.

MODELIZATION OF VISCOELASTIC BEHAVIOURS OF CARBON/EPOXY UNIDIRECTIONAL COMPOSITES

Yong-Sok O[*], C. PILLOT[+] and C. JOURDAN[+]

[*] Dow Pipe Systems, Tourville la Rivière, France.
[+] Institut National des Sciences Appliquées, Lyon, France.

ABSTRACT

This paper describes analytical and experimental efforts to characterize
and predict the viscoelastic properties of carbon fiber reinforced epoxy (CFRE)
unidirectional composites. The dynamic   mechanical behaviours are measured
by aid of a forced flexural vibration method in isothermal conditions with
the frequency range of 7.8 to 500 Hz. On the basis of time-temperature-
superposition principle, master curves are established. And then, through
a modelization of their dynamic behaviours, the master curves are reconstruc-
ted after having determined several characteristic parameters. The modelization, taking into account the dyssymetric responses of materials in time
and in temperature, can readily be converted into another viscoelastic functions showing the overall material performance and/or the differences in
fiber-matrix interface of composite materials.

INTRODUCTION

Despite the use of carbon fiber reinforced epoxy (CFRE) composites has
made great strides in many fields, the general understanding of such materials remains far from complete. Though many noteworthy research works
have been carried out on static and/or dynamic mechanical properties of
CFRE, few had been reported on the modelization of their viscoelastic properties.

We have previously developed a rapid, simple and accurate method to determine the dynamic mechanical properties of materials having high stiffness (1,2). In the present study the method is applied to characterize
first unfilled epoxy resins, and then their unidirectional carbon fiber
reinforced composites.

This paper represents the first to modelize the viscoelastic behaviour
of structural epoxy resins and their state-of-the-art composites. This
study is particularly important view of the growing applications of carbon/
epoxy composites as structural parts in aerospace and automotive industries.

EXPERIMENTAL

Materials
    The materials employed for this study are those widely used in aerospace
industries : NARMCO 5208 and CIBA-GEIGY REDUX 914. The NARMCO 5208 epoxy
resin is one of several commercial epoxies based upon the chemistry of

tetraglycidyl 4,4'-diamino-diphenyl methane (TGDDM or TGMDA) crosslinked by 4,4'-diamino-dipheny sulfone (DDS). The NARMCO 5208 system contains about 75,6 % by weight of TGDDM, about 19,7 % of DDS, and in addition about 4,7 % of diglycidyl ether of a bisphenol-A novolac epoxy.

As for CIBA-GEIGY 914, it contains about 27,6 % of TGDDM, about 41,4 % of triglycidyl ether of P-aminophenol, about 3,5 % of dicyanodiamide (DDA), and 27,6 % of polyether sulfone (PES).

Both epoxy resins were reinforced with 63 % by volume of TORAYCA T300B. The composite was fabricated from unidirectional prepreg and the number of plies is of 16, having the gauge thickness of 2.02 mm. Specimens with the dimension of 5.0 mm (width) x 60 mm (length) were prepared by dry cutting from a 300 mm x 300 mm panel.

For neat epoxy resin specimens, they were prepared by clear casting method. The gauge thickness was of 1.0 mm, and the width and length were, respectively, 10.0 mm and 60 mm.

Instrumentation
Figure 1 shows the schematic diagram of the test method : specimen holder and the corresponding mathematical formulae adapted to METRAVIB MAK-03 viscoelasticimeter.

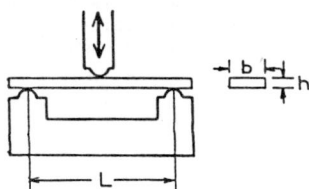

K : stiffness
δ : phase angle
f : frequency
M : mass of specimen
L : distance between two supports 1.50 mm
b : width, 5.0 mm
h : thickness, 2.02 mm

$$E' = \left[K \cdot \cos \delta + 2 \pi^2 f^2 M \right] \cdot \frac{24 \, L^3}{b \, h^3 \, \pi^4}$$

$$E'' = \left[K \cdot \sin \delta \right] \cdot \frac{24 \, L^3}{b \, h^3 \, \pi^4}$$

Fig. 1 Schematic diagram of specimen holder.

All the measurements were carried out within a linear viscoelastic region of materials (very small dynamic and static deformation).

All specimens were tested at as-received state without other thermal conditioning. Test temperature and frequency range are, respectively, from room temperature to about 300°C and from 7.8 Hz to 500 Hz.

RESULTS AND DISCUSSION

Neat resins
   Figure 2 shows the dynamic mechanical properties (real modulus and loss factor) of two neat epoxy resins as a function of temperature.

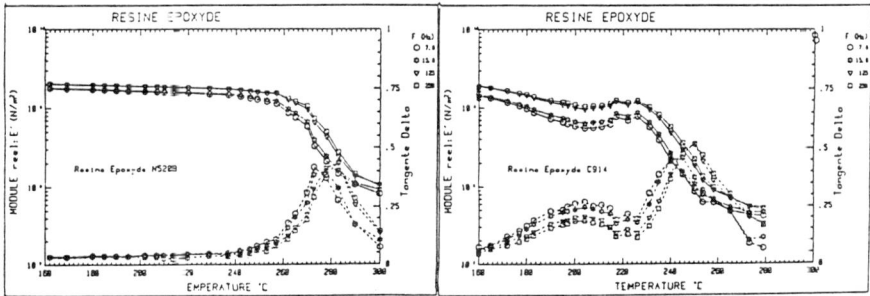

Fig. 2   Dynamic mechanical properties of neat epoxy resins.

   Though the real modulus of the two resins (N5208 and C914) are comparable each other at glassy state, the high temperature behaviours are quite different.   The C914 has more important damping properties than the N5208. Furthermore, even though both resins are submitted to a same curing cycle which is an industrial standard, the C914 shows unsufficient curing.

Unidirectional composites
   We have presented in Fig. 3 the dynamic mechanical properties of 4 specimens :

$$T\ 300/5208 \quad at \quad \theta = 0$$
$$T\ 300/914 \quad at \quad \theta = 0$$
$$T\ 300/5208 \quad at \quad \theta = 45$$
$$T\ 300/914 \quad at \quad \theta = 45$$

   One can note there is not a significant difference between the two composites of $\theta = 0$, only the properties of fibers dominate the overall behaviours of composites.   In the other hand, at the loading angle of $\theta = 45$, the contribution of matrix and, especially, that of fiber-matrix interface to composites' behaviours is easily to be detected.

   Thus, in Fig. 3, the comparison of the two composites at $\theta = 45$ indicates the T300/914 has higher viscoelastic behaviour than the T300/5208.   And such difference comes from the resin itself and, further, from the fiber-matrix interface.

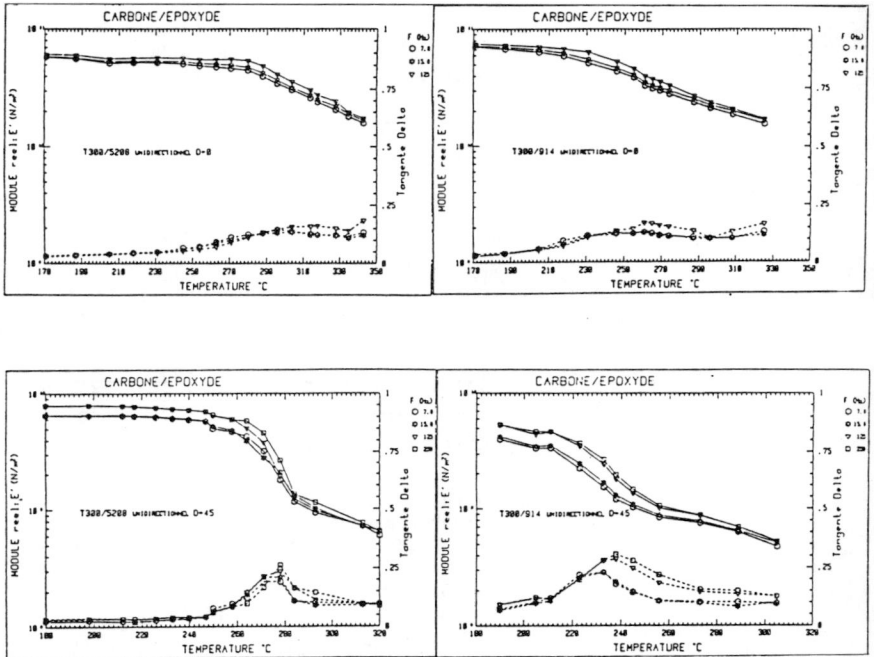

Fig. 3   Real modulus and loss factor versus temperature.

Master curves

   Master curves of these six materials are established by using the time-
temperature superposition principle, based on the measurements of dynamic
mechanical properties. We have chosen the maximum temperature of loss
factor peak as reference temperature, and shift factors are calculated from
real moduli. The calculation was effected by aid of a personal computer
with the software we have developped.

   Figure 4 shows the master curves of two neat epoxy resins and their corresponding unidirectional composites with loading angle θ = 0 and θ = 45.

   Presented in Fig. 5, the shift factor versus temperature plot of each
material shows a good fitting in the transition region.

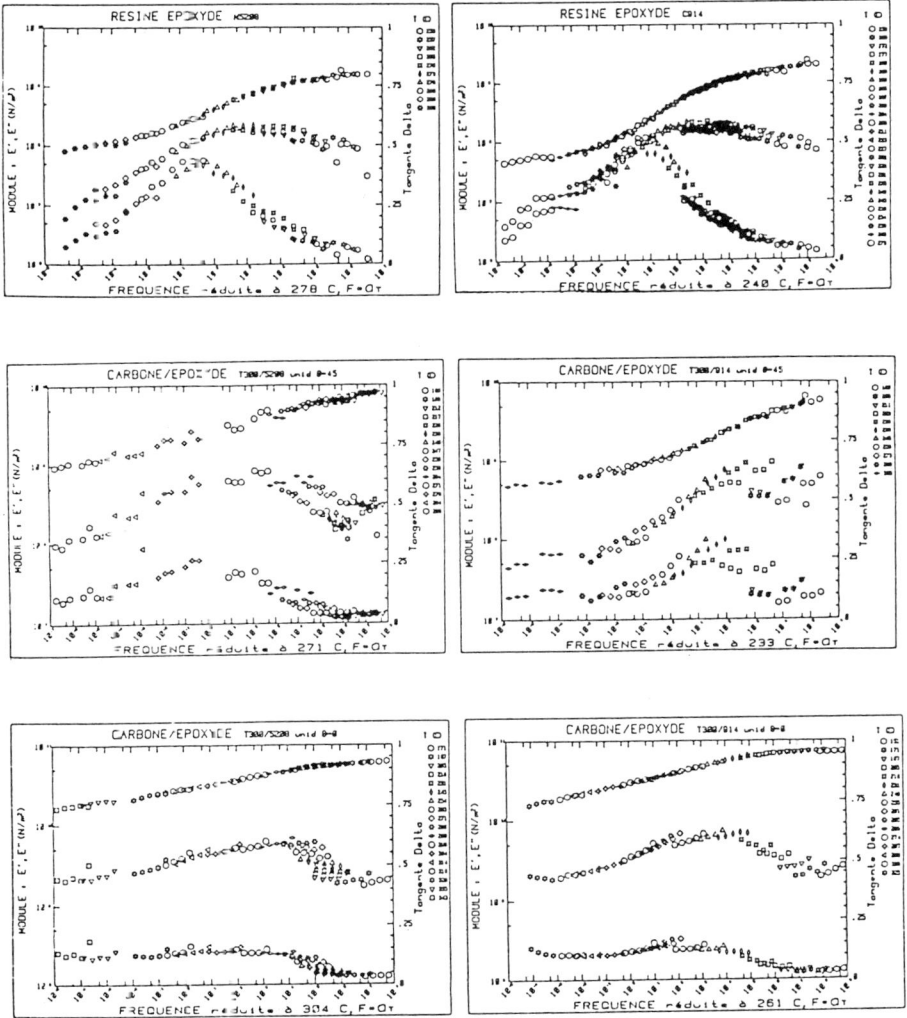

Fig. 4 Master curves of neat epoxy resins and their corresponding unidirectional composites at θ = 0 and θ = 45.

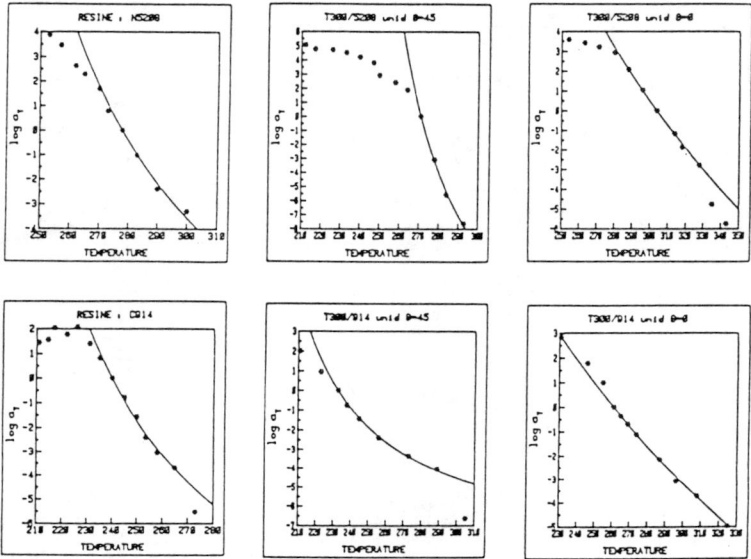

Fig. 5 Shift factor versus temperature.

Modelization

In order to establish a comparison and an identification the test results we have modelized permitting the interpretation of mechanical relaxation curves. The principal interest of the modelization is that, with limited number of parameters, it can easily visualize the behaviours of many materials.

We have chosen a mathematical model that describes almost perfectly the material behaviours. The model takes into account the materials' dyssymetric responses in time and in temperature.

Figure 6 shows well such dyssymetric responses when all the experimental points are presented in the form of Cole-Cole's diagram.

The mechanical model that describes most correctly such dynamic behaviours is shown in Fig. 7.

The mathematical expression of such model is :

$$E = E_o + \frac{E_\infty - E_o}{1 + \delta(1 \ \omega \ \tau)^{-h} + (1 \ \omega \ \tau)^{-k}}$$

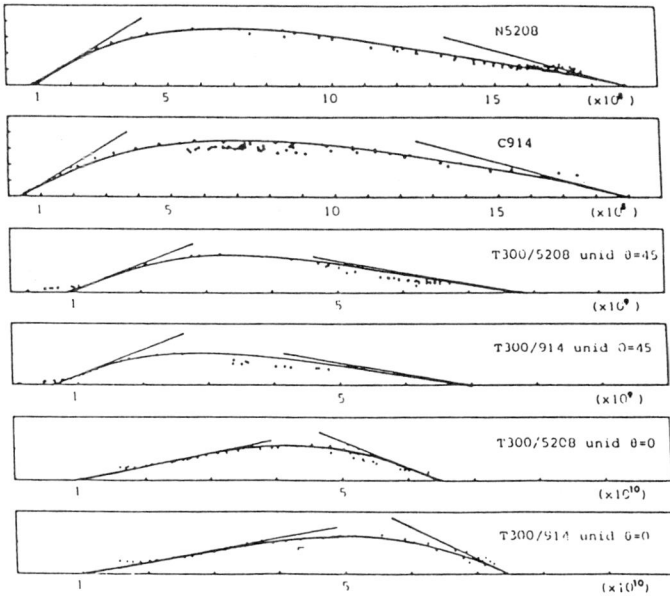

Fig. 6 Diagram of Cole-Cole.

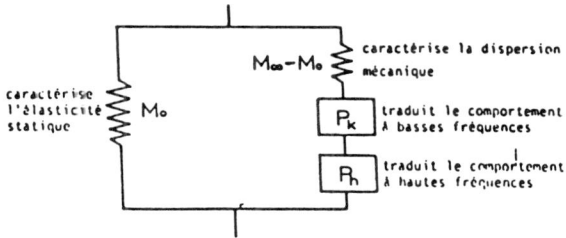

Fig. 7 Schematic representation of the model.

The determination of parameters can be effected through the method shown in Fig. 8, where

$$h = \alpha/(\pi/2)$$
$$\kappa = \beta/(\pi/2)$$
$$\delta = \frac{E_o - E_o}{\overline{AB}}$$

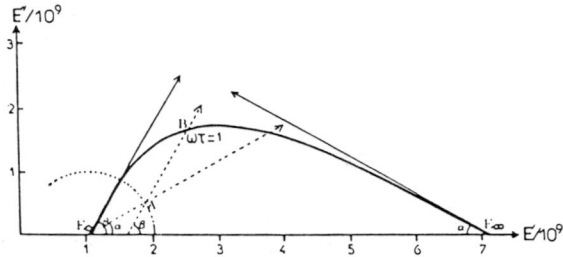

Fig. 8 Determination of parameters for parallel biparabolic viscoelastic model.

Parameters thus obtained are regrouped in Table 1.

Table 1

		δ	h	k	E_o (N/m^2)	E (N/m^2)
N 5208		6.01	0.17	0.30	8.6 10^7	1.93 10^9
C 914		6.98	0.17	0.36	5.24 10^7	1.92 10^9
T300/5208	=0	2.6	0.13	0.24	1.02 10^{10}	6.51 10^{10}
T300/5208	=45	2.7	0.10	0.17	8.98 10^8	8.02 10^9
T300/914	=0	2.58	0.12	0.30	1.08 10^{10}	7.55 10^{10}
T300/914	=45	3.84	0.13	0.26	5.06 10^8	7.01 10^9

One can note the presence of reinforcements lowers the relaxation process. The decrease of relaxation at long time for the system T300/5208 is more intensified than for the system T300/914. This results in an increase of viscoelasticity of T300/914. The increase in viscoelasticity can explain the absence of "blocking" of free chain ends by fiber surface.

Figure 9 shows the reconstruction of master curves through the modelization. It is shown that the theoretical curves fit well to the experimental points.

Real polymers do not have a single relaxation time, but a distribution of such time due to the polymolecularity, different side chains, and different relaxation processes. The presence of reinforcements and their concentration modify the distribution of relaxation time, since such presence induces morphological changes of matrix at interface.

The mathematical expression of the relaxation time spectra of the model can be written :

$$H_{(\tau)}/T=T_o = (E_\infty - E_o) \frac{[\delta h(\tau/\tau_o)^h + \kappa(\tau/\tau_o)^\kappa]}{[1 + (\tau/\tau_o)^h + (\tau/\tau_o)^\kappa]^2}$$

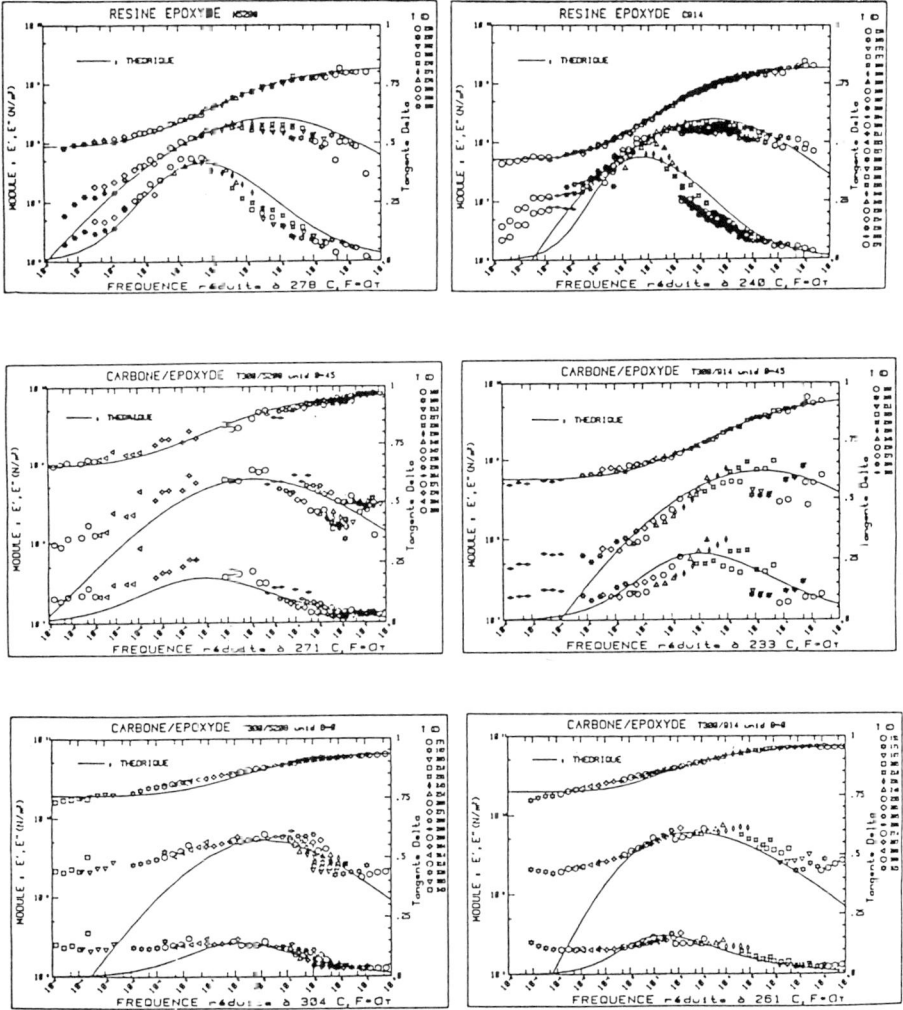

Fig. 9   Master curves and theoretical model.

Figure 10 illustrates the relaxation time spectra of neat epoxy resins and
their corresponding composites. It should be noted the reference temperature
of each specimen is different from one another. Thus, the curves can not be
interpreted in view of absolute relaxation time distribution at a given temperature, but they show the overall tendency of relaxation phenomena.

Neat resins present almost same relaxation time distribution. This is
true for their composite with θ = 0. In case of composites at = 45, the
conjugated effect of resin and interface on relaxation time distribution
becomes more complex. From the relaxation time spectra of composites at
θ = 45, one can conclude that the absorption of resin's free chain ends by
the surface of fibers is more efficient in T300/5208 than in T300/914.

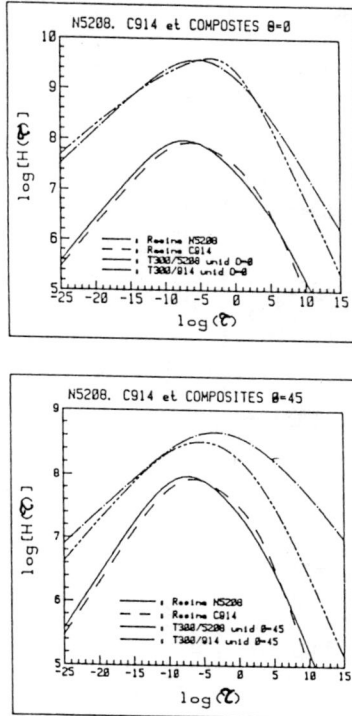

Fig. 10 Relaxation time spectra of neat epoxy resins and composites.

CONCLUSIONS

We have developed a fine analysis method of dynamic mechanical behaviours
of highly stiffened materials. This method helps the better characterization
of composites viscoelastic properties in relation with the morphology of
their matrix and of the fiber-matrix interface. Through the modelization
of rheological properties, the influence of the matrix or the interface on
the overall material performance can be more finely analysed. The modelization makes us to suggest that the presence of carbon fibers modify the
reaction mechanism of matrix, especially at the interface, by selective
absorption of free chain ends at fiber surface.

CHARACTERIZATION OF THE THERMOVISCOELASTIC BEHAVIOUR OF KEVLAR 49/EPOXY
LAMINATES

A. VAUTRIN

Ecole des Mines de Saint-Etienne, France.

The importance of fiber reinforced polymers as structural materials is
steadily growing. Polymers used as matrix materials exhibit behaviour
highly dependent on time and temperature. Thus, it is necessary to measure
the time and temperature effects upon the mechanical behaviour of fiber
reinforced polymers and to build up a reasonable anisotropic model.

The purpose of the present investigation is to characterize the time temperature mechanical behaviour of a unidirectional Kevlar 49/epoxy laminate
and to establish an anisotropic rheological model which could eventually
be used in composite structures design.

It is well known that the mechanical behaviour of polymer structures is
essentially viscoelastic. The time temperature superposition principle is
often used to characterize their behaviour in a given time temperature region (1). This procedure leads to the definition of shift factors which
provide a large amount of informations about the material microstructure
and can be helpful to develop accelerated characterization techniques (2).
Methods of reduced variables are also applicable with multiphase polymeric
systems and it must be emphasized that rough shift distances give insight
into the intricate relations between time and temperature dependance of
the mechanical characterization and modelization of the time temperature
responses of fiber reinforced polymers (4,5,6).

The mechanical response of the unidirectional laminate can be reasonably
modelized in an orthotropic framework. At the level of the infinitesimal
element of the macroscopic description, the relaxation function has an orthotropic symmetry. We admit that the symmetry is not modified by the initial
constrained configuration of reference and that the orthotropic axis are
invariants. The transverse isotropy in the 2,3 plane perpendicular to the
fibers direction 1 is not obvious because of the manufacturing of the laminate.

The isothermal, sinusoidal, quasi static, infinitesimal response in the
neighbourhood of a constrained mechanical configuration K_T, at the temperature T, is characterized by the complex compliances tensor $(S^\star)$:

$$(\varepsilon^\star) = (S^\star)\ (\sigma^\star)$$

where $(\varepsilon^\star)$ and $(\sigma^\star)$ are respectively the complex strain tensor and the complex stress tensor defined in K_T. In the case of thermorheologically simple
solid, $(S^\star)$ is function of the reduced frequency $Na(T,T_o)$ where N is the

frequency and a the shift factor.

The complex moduli tensor ($C^\star$) is naturally expressed with the relaxation functions and the complex compliances tensor is the inverse of the complex moduli tensor.

If we limit our investigation to the 1, 2 plane, we get in contracted notations :

$$(S^\star) = \begin{vmatrix} S^\star_{11}[\,Na] & S^\star_{12}[\,Na] & 0 \\ S^\star_{21}[\,Na] & S^\star_{22}[\,Na] & 0 \\ 0 & 0 & S^\star_{66}[\,Na] \end{vmatrix}$$

Fig. 1   Principal complex compliance at a given test temperature, N is the frequency, Na is the reduced frequency and 1 is the fiber direction.

It is advisable to define the phase angle of the viscoelastic strains with respect to the elastic strains ; so the dynamic compliances matrix can be given by :

$$\begin{vmatrix} S_{11} & S_{12} & 0 \\ S_{21} & S_{22} & 0 \\ 0 & 0 & S_{66} \end{vmatrix} \qquad \text{where} \qquad \begin{aligned} S_{ii} &= \|S^\star_{ii}\| \\ S_{ij} &= -\|S^\star_{ij}\| \end{aligned}$$

Fig. 2   Dynamic principal compliances ;
$\|S^\star_{ij}\|$ is the modulus of the complex compliance S_{ij}.

When the phase angles are small, we can consider that the dynamic compliances reasonably approximate the real part of the complex compliances.

A tensile testing programe was conducted to measure and characterize the frequency temperature orthotropic response of a unidirectional Kevlar 49/epoxy (1) laminate ($v^f = 0,60$) in the [$5.10^{-4} - 5.10^{-2}$ Hz, 200-343 K] range (7).

Four kinds of samples have been tested, each kind being defined by the value of the angle θ between the fibers direction 1 and the tension axis I. The values of the experimental parameters and of the measured and calculated functions have been reported in Table 1.
The strains are measured in three directions a, b, c respectively at $\beta = 0$, $\pi/4$ and $\pi/2$ with regard to the axis of uniaxial tension I for each θ ; they are noted $\varepsilon(\theta;\beta)$. The associated experimental dynamic compliances are :

$$J(\theta;\beta) = \frac{\varepsilon(\theta;\beta)}{\sigma_I}$$

Table 1 Samples, experimental parameters and calculated compliances.

	$\theta = 0$	$\theta = -\frac{\pi}{16}$	$\theta = -\frac{\pi}{4}$	$\theta = -\frac{\pi}{2}$
Test temperatures (K)	215,7 232,9 252,4 267,7 282,9 298,5 316,9 332,4	202,2 227,5 243,3 257,8 267,4 276,2 288,3 298,9 318 342,5	201,7 228,9 246 260,2 268,6 278,7 288,7 298,2 307,1 317,9 333,5	201,3 224,1 243,3 259,1 267,7 277,7 289 298,2 313,7 327,7 343,6
Measured strains and phase angles	$\sigma_a = \sigma_I$ $\varepsilon_a = \varepsilon(0;0)$ $\varepsilon_b = \varepsilon(0;\frac{\pi}{4})$ $\varepsilon_c = \varepsilon(0;\frac{\pi}{2})$ $\phi_a = \phi_{II}$	$\sigma_a = \sigma_I$ $\varepsilon_a = \varepsilon(-\frac{\pi}{16};0)$ $\varepsilon_b = \varepsilon(-\frac{\pi}{16};\frac{\pi}{4})$ $\varepsilon_c = \varepsilon(-\frac{\pi}{16};\frac{\pi}{2})$ $\phi_a = \phi_{II}$	$\sigma_a = \sigma_I$ $\varepsilon_a = \varepsilon(-\frac{\pi}{4};0)$ $\varepsilon_b = \varepsilon(-\frac{\pi}{4};\frac{\pi}{4})$ $\varepsilon_c = \varepsilon(-\frac{\pi}{4};\frac{\pi}{2})$ $\phi_a = \phi_{II}$	$\sigma_a = \sigma_I$ $\varepsilon_a = \varepsilon(-\frac{\pi}{2};0)$ $\varepsilon_b = \varepsilon(-\frac{\pi}{2};\frac{\pi}{4})$ $\varepsilon_c = \varepsilon(-\frac{\pi}{2};\frac{\pi}{2})$ $\phi_a = \phi_{II}$
Calculated quantities	$J(0;0) = J_{II}$ $J(0;\frac{\pi}{4})$ $J(0;\frac{\pi}{2}) = J_{II\,I}$ $tg\,\phi_{II}$	$J(-\frac{\pi}{16};0) = J_{II}$ $J(-\frac{\pi}{16};\frac{\pi}{4})$ $J(-\frac{\pi}{16};\frac{\pi}{2}) = J_{II\,I}$ $tg\,\phi_{II}$	$J(-\frac{\pi}{4};0) = J_{II}$ $J(-\frac{\pi}{4};\frac{\pi}{4})$ $J(-\frac{\pi}{4};\frac{\pi}{2}) = J_{II\,I}$ $tg\,\phi_{II}$	$J(-\frac{\pi}{2};0) = J_{II}$ $J(-\frac{\pi}{2};\frac{\pi}{4})$ $J(-\frac{\pi}{2};\frac{\pi}{2}) = J_{II\,I}$ $tg\,\phi_{II}$

where σ_I is the amplitude of the uniaxial sinusoidal stress. The phase angle Φ_{II} between the uniaxial imposed stress and the strain $\varepsilon(0;0)$ has only been measured (Table 1) and gives rise to the damping ratio $tg\ \Phi_{II}$.

The modelization of the behaviour uses all the experimental informations which have been collected. It needs the definition of a reference mechanical configuration independent of the different tests and requires the knowledge of all the values of the compliances at the same temperatures. The experiments do not allow us to take into account the effects of the internal stresses rigorously. Nevertheless it is quite possible to deduce the compliances at different conventional temperatures which are common to the four groups of samples. We have adopted the following original way for the analysis of the results :

1) Corrections of the temperature and transverse effects on the responses of the strain gages.

2) Construction of the variations at fixed frequency N of each compliance versus the test temperature T (Figs. 3 and 4).

3) Smoothing of the preceding evolutions (Figs. 3 and 4).

4) Smoothed compliances at the following conventional temperatures : $T_n = 90 + 10\ n$ (°K) where $n = 1,2,...,16$.

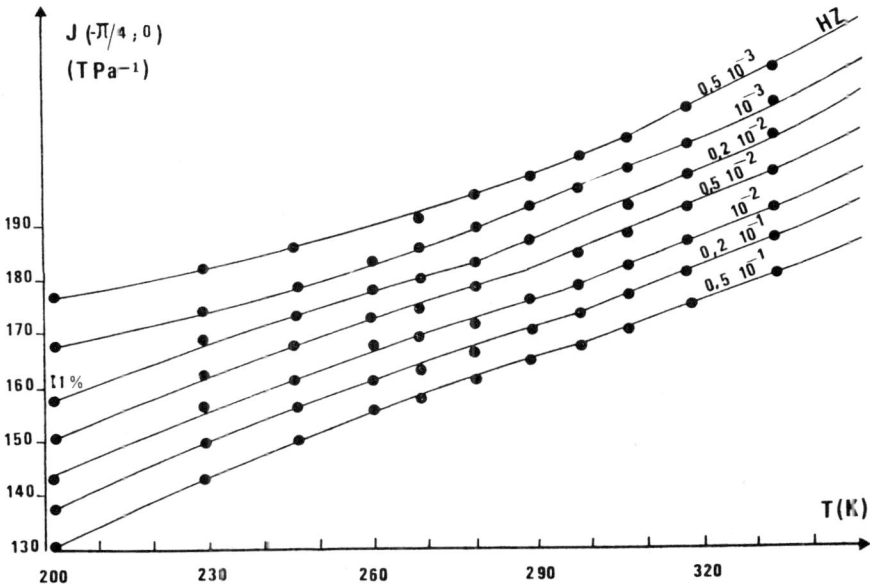

Fig. 3 T-N network for $J(-\pi/4;0)$, (experimental points and smoothed curves).

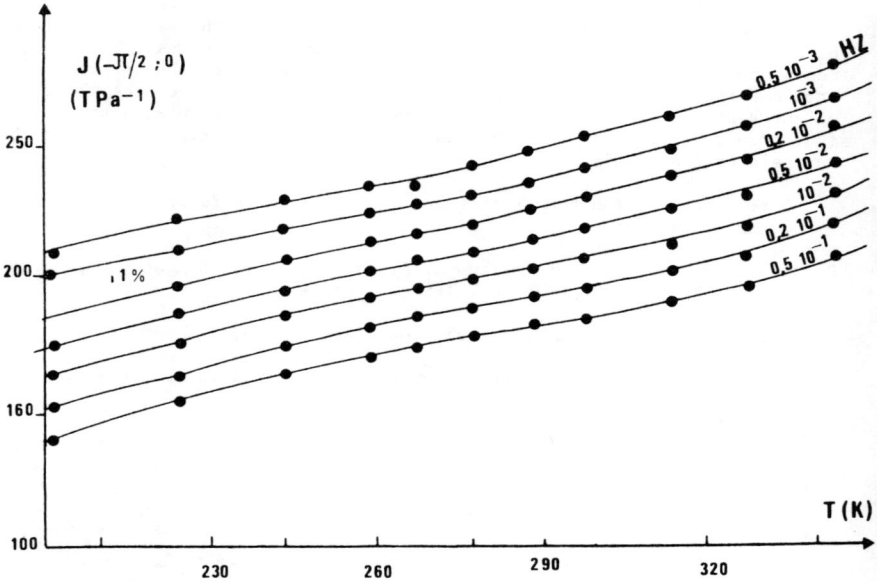

Fig. 4   T-N network for $J(-\pi/2;0)$, (experimental points and smoothed curves).

N.B. : For more clearness the different consecutive curves have been vertically shifted of 5 TPa^{-1} (Fig. 3) and 10 TPa^{-1} (Fig. 4) from the curve
at $N = 0.5\ 10^{-1}$ Hz.

So we finally got 120 tables of corrected dynamic compliances such like
Table 2.

It must be emphasized that this procedure will considerably reduce the
effects of measurement errors on the mechanical description of the behaviour.
The reduction of these errors effects will permit future severe checks of
the coherence of the model. This point is quite fundamental because the
purpose of the study is to build up a reasonnable description in a large
Time-Temperature range.

Table 2   Corrected dynamic compliances at $(N,T) = (0.2 \ 10^{-2}$ Hz, $298 \pm 1K)$
in TPa^{-1}.

θ \ β	0	π/4	π/2
0	14,1	4,16	-6,05
-π/16	31,4	-35,4	-17,9
-π/4	170	7,97	-82,3
-π/2	190	93,3	-9,78

The corrected data can now be used to determine the principal dynamic
compliance for each (N,T) couple. So the problem is to estimate four quantities from the twelve corrected dynamic compliances. This problem is obviously redundant. We used an identification least – squares method to
compute at best the principal compliances (8,9). At last it was possible
to plot the variations of these final quantities versus the frequency at
the different conventional temperatures ; we present here the evolution of
S_{11} (Fig. 5) and S_{2} (Fig. 6).

We checked the coherence of our results by looking at the differences
between the experimental data and the values which can be calculated from
the optimized principal dynamic compliances. As it appear that the calculated evaluations lay within the data range, we concluded that the analysis
was consistent.

So the four components of the dynamic law are frequency and temperature
dependents. Moreover it seems that a linear viscoelastic model would be an
acceptable estimate. In particular it leads to a good prediction of the
experimental evolutions of the damping ratios versus N.

The time temperature superposition principle is then used to estimate at
best the master curve of every principal dynamic compliance and deduce the
associated shift factors (the calculation has been done on a HP 85 microcomputer). The master curves fit the experimental data very well and fair
agreement is obtained between the resulting predictions outside of the experimental range and the direct measurement of the dynamic compliances by
ultrasonor wave propagation.

It is well known that the building of the master curve depends upon the
arrangement of the different isothermal curves. In the actual case we got
a very reasonable precision on the shiftings because of our preceding
choice of conventional temperatures which provided a good overlapping of
the consecutive curves.

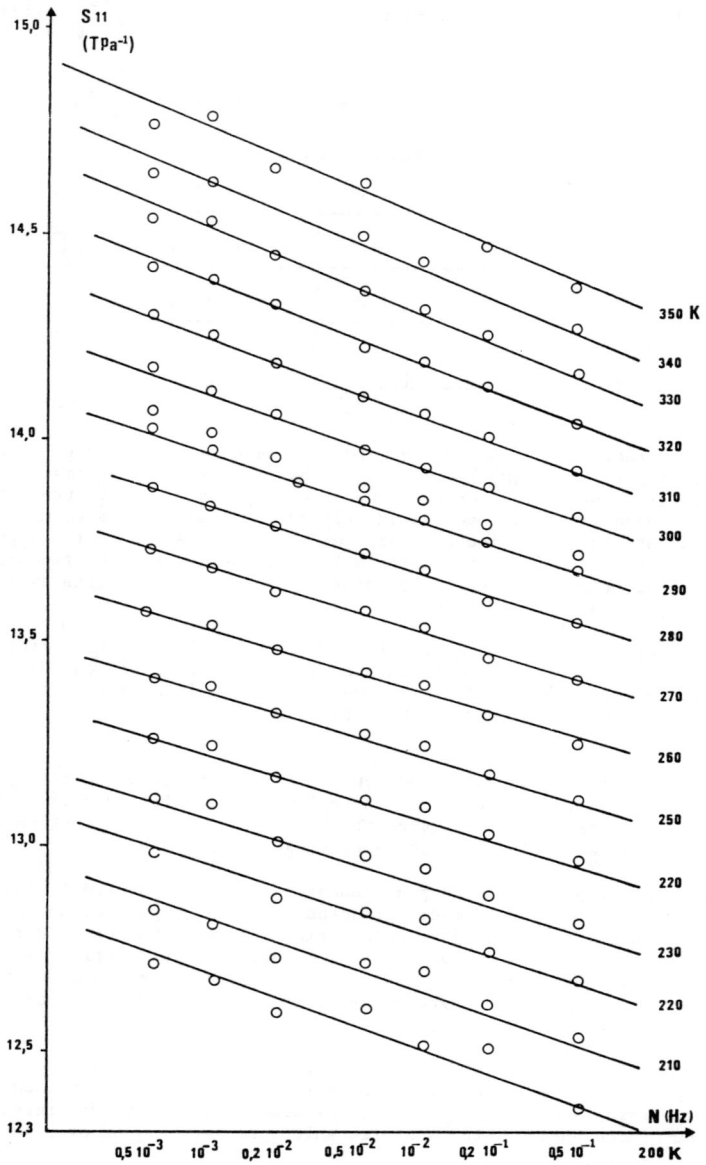

Fig. 5 N–T network and isothermal master curve for S_{11}.

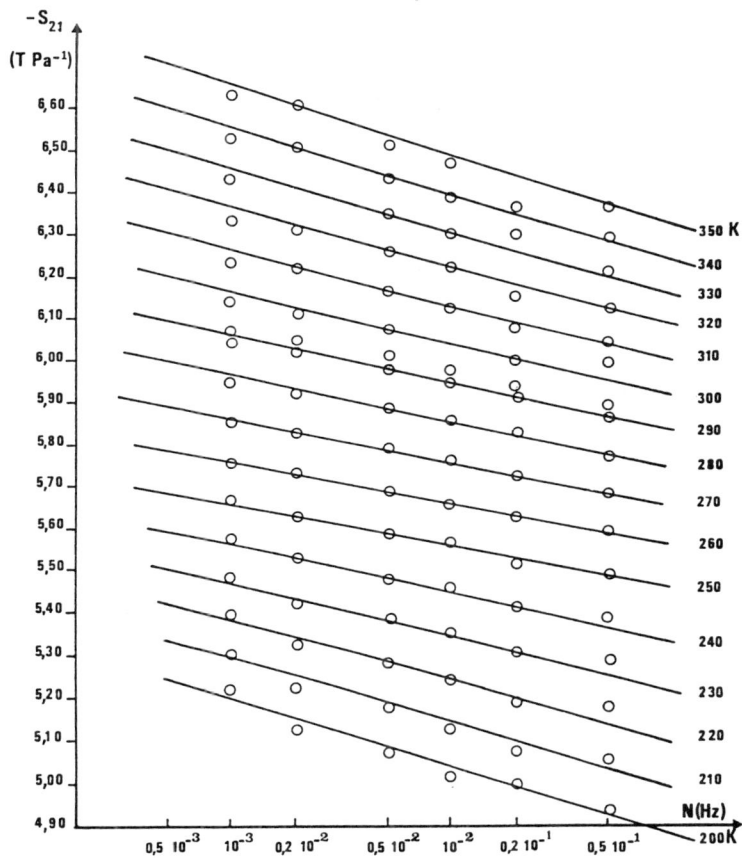

Fig. 6 N-T network and isothermal master curves for S_{21}.

This reduction clearly establishes that two different shift factors a^f and a^m (Fig. 7) are needed to characterize the frequency temperature viscoelastic response of the material.

$$(S^\star) = \begin{vmatrix} S^\star_{11}[Na^f] & S^\star_{12}[Na^f] & 0 \\ S^\star_{21}[Na^f] & S^\star_{22}[Na^m] & 0 \\ 0 & 0 & S^\star_{66}[Na^m] \end{vmatrix}$$

Fig. 7 Anisotropic model for the frequency-temperature dependence of the complex compliances.

So the shift factor looses its intrinsic nature and we must notice that the isothermal curves of the off-axis compliances cannot, in principle, be brought into superposition by a simple shift along the logarithmic frequency axis because the shift distances become function of frequency in addition to temperature.
At this point, the orthotropic frequency temperature behaviour of the composite could be characterized by six functions :

- four principal compliances functions of the frequency at a given reference temperature (Fig. 8)
- two shift factors a^f and a^m functions of the test temperature (Fig. 9).

These two experimental parameters a^f and a^m are respectively specific of the properties of the fiber or the interface fiber/matrix and of the matrix alone. We experimentally verified that a^m is the shift factor of the matrix in the frequency temperature range of interest. As we could not carry out or find in the literature any reliable experiments on the fibers alone, it was impossible to determine the origin of a^f more precisely.

The frequency temperature behaviour which has been detected is clearly induced by the constituent (the fiber, the interface, the matrix) the contribution of which is predominant in the deformation process. This result is fairly intuitive of course, but is conceals an important aspect of the true behaviour of the composite because it means that the parts of the constituents are balanced. The following analysis of the net curves of the off-axis compliances will show that the time temperature behaviour of the composite is essentially due to the matrix.

The determination of the whole set of four principal compliances at (N,T) is quite independent of every time temperature model, so these net curves can be used to check the validity of any new estimation of the off-axis compliances. Assuming the orthotropic material symmetry, we derive the isothermal master curves of fourteen distinct off-axis compliances from two different approaches :

1) Computation of the net curves of the off-axis compliance from the net curves of the principal compliances and construction of the master curve at a given reference temperature according to the well known shifting procedure.

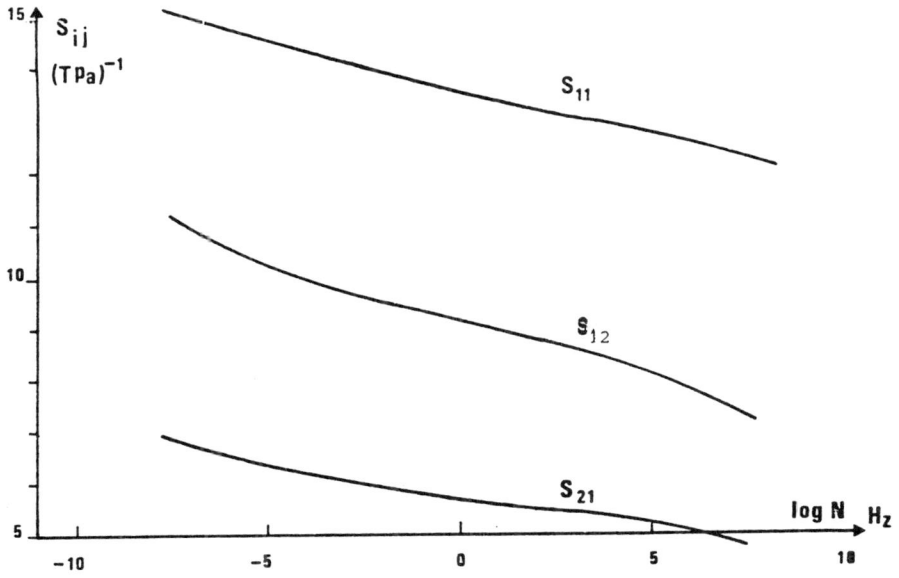

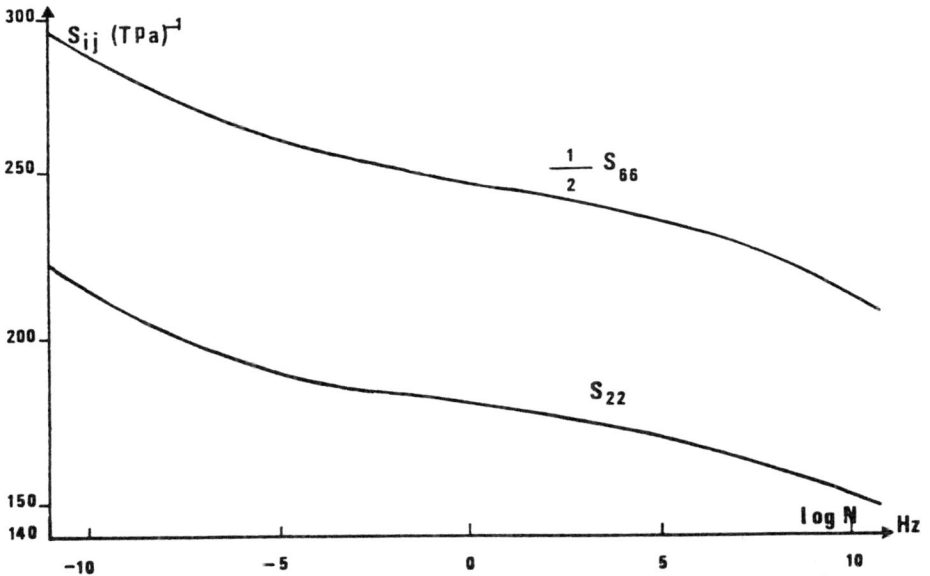

Fig. 8 Modulus of the principal compliances versus the frequency N at the reference temperature $T_o = 293$ K.

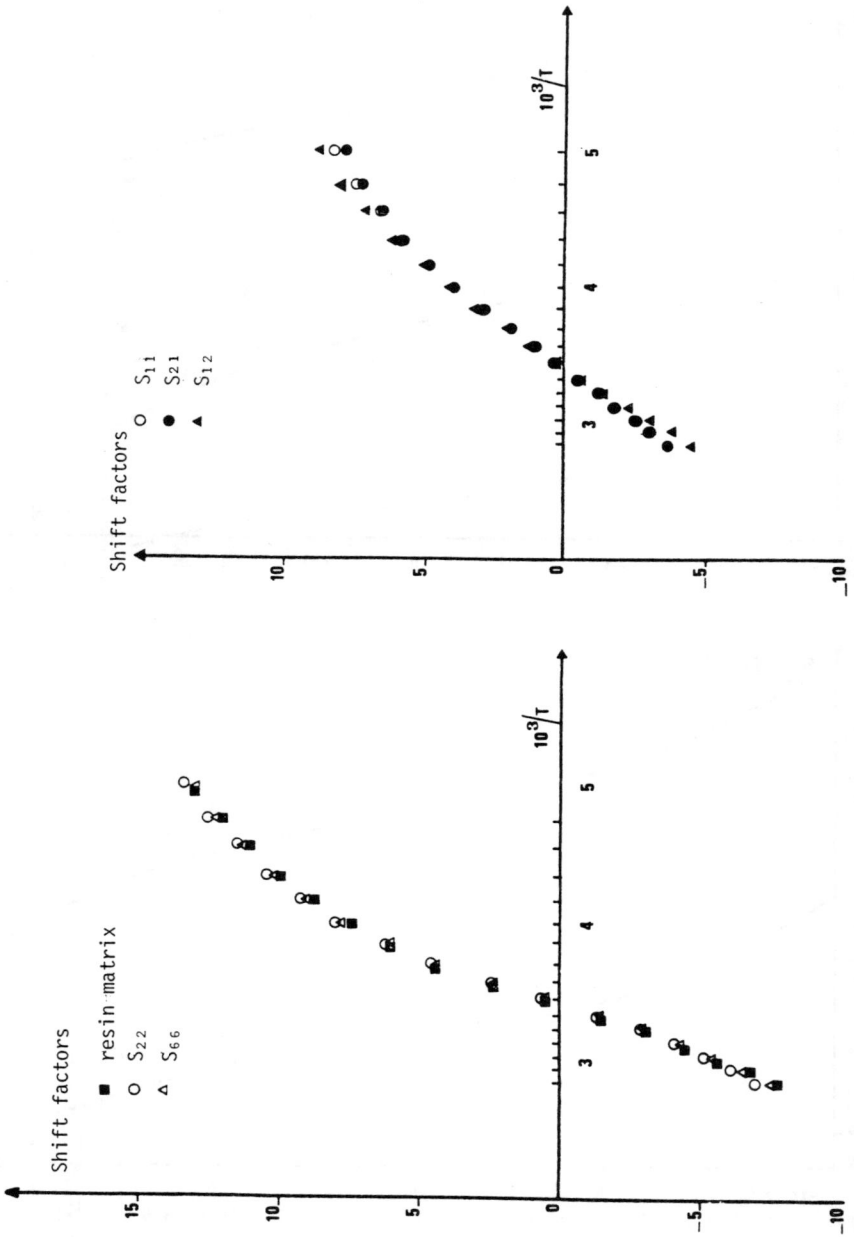

Fig. 9 Experimental shift factors deduced of the construction of the different master curves.

2) Direct computation of the master curve from the principal compliance
master curves previously constructed.

Three different reference temperatures have been used. It clearly appears
that the time temperature superposition principle can lead to a good approximation of the frequency evolution of the off-axis compliances in the
experimental range although the rheological model involves two distinct shift
factors. A reliable way of computation of the off-axis shift factor is given.
Furthermore, the off-axis shift factors reduce to the matrix shift factor a^m
when the fiber orientation angle is in the [30° - 60°] range.

So the study of the reduction of the off-axis compliances net curves
leads to :

1) A better knowledge of the deformation mechanisms of the fiber reinforced
polymer.

2) A new phenomenological approached model which can be easily used to
compute the mechanical response of laminates in a large frequency and
temperature range.

The validity of the prediction for laminates must be carefully checked
especially when :

1) The fiber angles are in the vicinity of 0° and 90°.

2) The laminates are subject to complex state of stresses. It is likely
that the prediction when the fiber angles are within the [30° - 60°]
range will be quite reasonable because of the influence of the matrix
but this point has not been yet verified.

REFERENCES

1. Schapery, R.A., Viscoelastic Behaviour and Analysis of Composite
Materials.
Composite Materials, vol. 2, pp. 86-165, G.P. Sendeckyj, Academic Press,
New York, 1974.

2. Jones, D.I.G., Temperature-Frequency Dependence of Dynamic Properties
of Damping Materials.
Journal of Sound and Vibration, 33, 4, pp. 451-470, 1974.

3. Kaplan, D. and Tschoegl, N.W., Time Temperature Superposition in Two-
Phase Polyblends.
Polymer Engineering and Science, 14, 1, pp. 43-49, 1974.

4. Moehlenpah, A.E., Ishai, O. and Dibenedetto, A.T., The Effect of Time
and Temperature on the Mechanical Behaviour of Epoxy Composites.
Polymer Engineering and Science, 11, 2, pp. 129-138, 1971.

5. Reed, K.E., Dynamic Mechanical Analysis of Fiber Reinforced Composites.
Polymer Composites, 1, 1, pp. 44-49, 1980.

6. Morris, D.H., Brinson, H.F. and Yeow, Y.T., The Viscoelastic Behaviour of the Principal Compliance Matrix of a Unidirectional Graphite/Epoxy Composite.
   Polymer Composites, 1, 1, pp. 32-36, 1980.

7. Vautrin, A., Contribution à la Caractérisation Mécanique des Composites à Renfort Filamentaire.
   Thèse de doctorat ès sciences, Nancy, 1983.

8. Vong, T.S. and Verchery, G., Optimal Use of Redundant Measurements of Constrainted Quantities, Application to Elastic Moduli of Anisotropic Composite Materials.
   Proceeding of ICCM3, 2, Paris, pp. 1783-1795, 1980.

9. Vautrin, A., Estimation au mieux de la Loi Plane de Comportement d'une Plaque Orthotrope.
   Comptes Rendus des Quatrièmes Journées Nationales sur les Composites, Pluralis, Paris, pp. 211-228, 1984.

CRACK GROWTH DIRECTION IN UNIDIRECTIONAL OFF-AXIS GRAPHITE-EPOXY

C.T. HERAKOVICH, M A. GREGORY and J.L. BEUTH, Jr.

Virginia Polytechnic Institute and State University, Blacksburg, U.S.A.

ABSTRACT

An anisotropic elasticity crack tip stress analysis is implemented using three crack extension direction criteria (the normal stress ratio, the tensor polynomial and the strain energy density) to predict the direction of crack extension in unidirectional off-axis graphite-epoxy. The theoretical predictions of crack extension direction are then compared with experimental results for 15° off-axis tensile coupons with center cracks. Specimens of various aspect ratios and crack orientations are analyzed. It is shown that only the normal stress ratio criterion predicts the correct direction of crack growth.

INTRODUCTION

A fundamental problem in predicting the failure of laminated composite materials is prediction of the direction of crack growth in the individual laminae, and the laminate. The influence of the direction of crack growth on the failure response of the laminate is shown in fig.1, (1). The clustered $[\theta_2/-\theta_2]_s$ graphite-epoxy laminates fail in a pure matrix mode (delamination and either intralaminar matrix cracking or fiber matrix debonding). In contrast, the alternating $[(+\theta/-\theta)_2]_s$ laminates exhibit fiber breakage in half of the plies, and either matrix cracking or fiber matrix debonding in the others; there is no delamination. The mode of failure has a significant effect on the strength of the laminate. The strength of 10° and 30° alternating laminates is, for example, 30 and 50 percent greater, respectively, than the strength of clustered 10° and 30° laminates, (1). Hence, understanding the parameters that affect laminate failure, particularly those influencing the direction of crack growth in the lamina and between laminae, is of critical importance in predicting the fracture response of laminates.

Predicting the direction of crack extension in laminates is a very complex three-dimensional problem. Since the lamina is the basic building block of the laminate, its behavior must be fully understood as a stepping stone toward understanding the behavior of the laminate. This study was undertaken to assess more critically the applicability of three criteria which have been presented in the literature for predicting the direction of crack growth in unidirectional fibrous composites, (2-4).

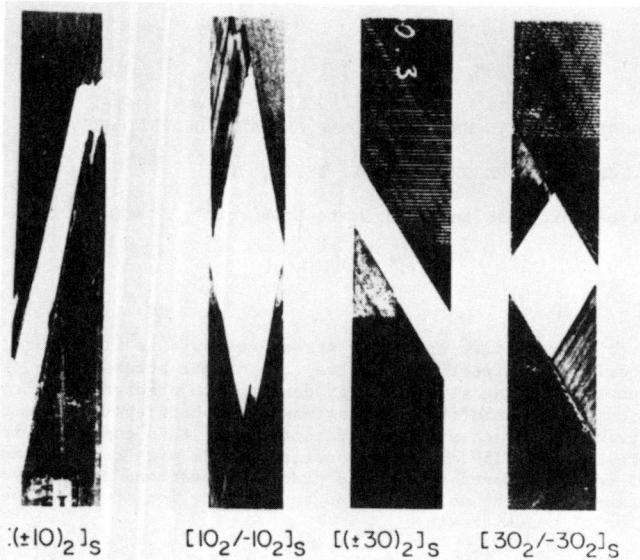

Fig. 1 Failure modes in angle-ply laminates.

CRACK EXTENSION DIRECTION CRITERIA

Three phenomenological criteria for predicting the direction of crack extension in homogeneous, anisotropic materials are the normal stress ratio criterion, (2), the tensor polynomial criterion (3) and the strain energy density criterion (4). These criteria can be used to predict the load at failure and the direction of crack extension. The crack tip coordinate system used in this analysis is shown in Fig. 2.

Normal stress ratio criterion

Buczek and Herakovich (2) have hypothesized the normal stress ratio criterion as a crack growth direction criterion. The model assumes that the direction of crack extension corresponds to the direction of the maximum value of the normal stress ratio $R(r_o,)$ where

$$R(r_o, \phi) = \frac{\sigma_{\phi\phi}}{T_{\phi\phi}} \tag{1}$$

In the expression for $R(r_o, \phi)$, $\sigma_{\phi\phi}$ corresponds to the normal stress acting on the radial plane defined by ϕ, and at a given distance, r_o, from the crack tip. $T_{\phi\phi}$ is the tensile strength on the ϕ plane.

Since the tensile strength on an arbitrary plane is difficult, if not impossible, to measure, $T_{\phi\phi}$ is defined in a manner consistent with the tests that can be performed.

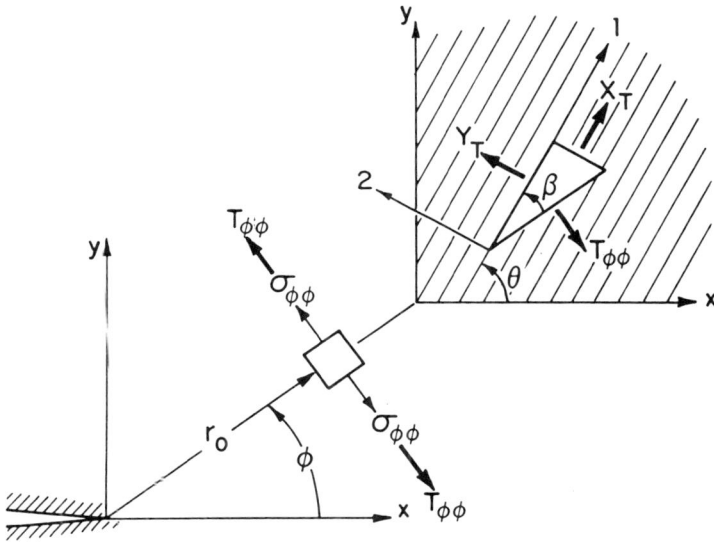

Fig. 2 Coordinate system and normal stress ratio parameters.

To meet this requirement, a mathematical definition of $T_{\phi\phi}$ must satisfy the following conditions :

1) for an isotropic material, $T_{\phi\phi}$ must be independent of ϕ ;

2) for crack growth parallel to the fibers, $T_{\phi\phi}$ must equal the transverse tensile strength Y_T ;

3) for crack growth perpendicular to the fibers, $T_{\phi\phi}$ must equal the longitudinal tensile strength X_T.

A definition satisfying these conditions is :

$$T_{\phi\phi} = X_T \sin^2\beta + Y_T \cos^2\beta \qquad (2)$$

where β is the angle from the plane of interest to the fiber direction.

Tensor polynomial criterion
    Tsai and Wu (5) first presented the tensor polynomial criterion as an anisotropic failure criterion. This criterion is based on the existence of a failure surface in stress space of the form :

$$F(\sigma_i) = F_i\sigma_i + F_{ij}\,\sigma_i\sigma_j \qquad (3)$$

where F_i and F_{ij} are strength tensors of second and fourth order, and σ_i

is the contracted form of the stress tensor. Expressions for F_i and F_{ij} are given in Table 1.

Table 1   Relationships for strength tensors in terms of measured strengths

$$F_1 = (1/X_T + 1/X_c)$$

$$F_2 = (1/Y_T + 1/Y_c)$$

$$F_6 = 0.0$$

$$F_{11} = -1/(X_T X_c)$$

$$F_{22} = -1/(Y_T Y_c)$$

$$F_{66} = 1/(S^2)$$

In application of the tensor polynomial to fracture problems (3), the assumed direction of crack extension is the radial direction of maximum $f(\sigma_i)$. The stress components σ_i are those determined by a continuum mechanics-based stress analysis, and must be evaluated at a finite distance, r_o, from the crack tip.

Strain energy density criterion
    The strain energy density criterion is based on variations in the energy stored along the periphery of a core region surrounding the crack. Sih presents the criterion for isotropic failure in (6) and a modified form for application to anisotropic fracture in (4).

    The strain energy density factor, S, is defined as :

$$\frac{\partial W}{\partial V} = \frac{S}{r} \tag{4}$$

where W/ V is the strain energy density function and r is the distance from the crack tip. Since the strain energy density function can be expressed in terms of the crack tip stresses and strains for plane stress as :

$$\frac{\partial W}{\partial V} = \frac{1}{2} (\sigma_x \varepsilon_x + \sigma_y \varepsilon_y + \tau_{xy} \gamma_{xy}) \tag{5}$$

an expression for the strain energy density factor, S, can be obtained by substitution. The resulting expression is :

$$S = \frac{r}{2} (\sigma_x \varepsilon_x + \sigma_y \varepsilon_y + \tau_{xy} \gamma_{xy}) \tag{6}$$

The fundamental hypothesis of Sih (4) for unstable crack growth is that crack initation takes place in the radial direction corresponding to a minimum value of the strain energy density factor, i.e.,

$$\frac{\partial S}{\partial \phi} = 0 \quad \text{and} \quad \frac{\partial^2 S}{\partial \phi^2} > 0 \qquad \text{at} \quad \phi = \phi_c \tag{7}$$

Sih cautions that for small values of r, a continuum mechanics-based crack tip stress analysis is invalid. Hence, the strain energy factor should be evaluated at a finite distance, r_0, from the crack tip, where r_0 is of the same order of magnitude as the crack tip curvature.

ANISOTROPIC ELASTICITY ANALYSIS OF CRACK TIP STRESS FIELDS

The stress analysis of an inifinite homogeneous anisotropic plate containing a center crack can be directly related to a homogeneous anisotropic plate with an elliptic hole. By reducing the minor axis dimension to zero and evaluating the stress potential functions in the neighborhood of the crack tip, Lekhnitskii's complex variable solution (7) for an elliptic hole in an anisotropic plate can be adapted to solve anisotropic fracture problems. Wu presents a detailed description of this procedure in (8), along with equations describing the crack tip stresses for an infinite homogeneous anistropic center-cracked plate. The problem under consideration is shown in Fig. 3.

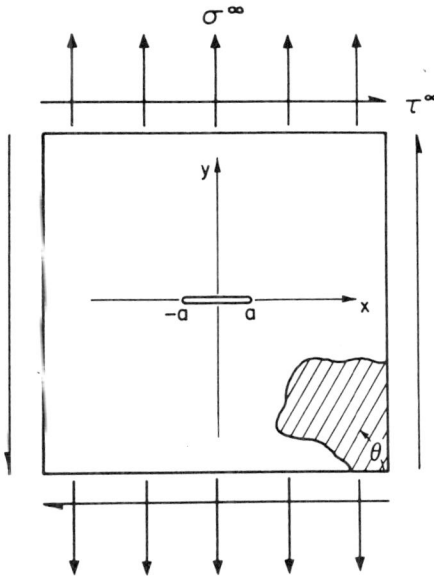

Fig. 3 Infinite center cracked plate under biaxial load.

The governing partial differential equation for this problem in terms of the Airy's stress function U is :

$$\frac{\partial^4 U}{\partial x^4} - \frac{2A_{26}}{A_{22}} \frac{\partial^4 U}{\partial x^3 \partial y} + \frac{(2A_{12} + A_{66})}{A_{22}} \frac{\partial^4 U}{\partial x^2 \partial y^2} - \frac{2A_{16}}{A_{22}} \frac{\partial^4 U}{\partial x \partial y^3} + \frac{A_{11}}{A_{22}} \frac{\partial^4 U}{\partial y^4} = 0 \qquad (8)$$

where A_{ij} are components of the compliance tensor for plane stress or plane strain, depending on the analysis desired.

Assuming $U = e^{x+sy}$, the characteristic equation for (8) takes the form :

$$A_{11}S^4 - 2A_{16}S^3 + (2A_{12} + A_{66})S^2 - 2A_{26}S + A_{22} = 0 \qquad (9)$$

The roots of the characteristic equation, S_1 and S_2, (and their conjugates) are complex, and are functions of the material properties and the orientation of the crack relative to the principal material direction.

Assuming $S_1 \neq S_2$, evaluation of the complex potential functions near the crack tip yields expressions for the stress and displacement distributions of the form :

$$\sigma_x = \frac{\sigma^\infty \sqrt{a}}{\sqrt{2r}} \, \mathrm{Re} \, \{ \frac{S_1 S_2}{(S_1 - S_2)} [\frac{S_2}{\psi_2^{1/2}} - \frac{S_1}{\psi_1^{1/2}}] \} + \frac{\tau^\infty \sqrt{a}}{\sqrt{2r}} \, \mathrm{Re} \, \{ \frac{1}{(S_1 - S_2)} [\frac{S_2^2}{\psi_2^{1/2}} - \frac{S_1^2}{\psi_1^{1/2}}] \}$$

$$\sigma_y = \frac{\sigma^\infty \sqrt{a}}{\sqrt{2r}} \, \mathrm{Re} \, \{ \frac{1}{(S_1 - S_2)} [\frac{S_1}{\psi_2^{1/2}} - \frac{S_2}{\psi_1^{1/2}}] \} + \frac{\tau^\infty \sqrt{a}}{\sqrt{2r}} \, \mathrm{Re} \, \{ \frac{1}{(S_1 - S_2)} [\frac{1}{\psi_2^{1/2}} - \frac{1}{\psi_1^{1/2}}] \}$$

$$\tau_{xy} = \frac{\sigma^\infty \sqrt{a}}{\sqrt{2r}} \, \mathrm{Re} \, \{ \frac{S_1 S_2}{(S_1 - S_2)} [\frac{1}{\psi_1^{1/2}} - \frac{1}{\psi_2^{1/2}}] \} + \frac{\tau^\infty \sqrt{a}}{\sqrt{2r}} \, \mathrm{Re} \, \{ \frac{1}{(S_1 - S_2)} [\frac{S_1}{\psi_1^{1/2}} - \frac{S_2}{\psi_2^{1/2}}] \}$$

$$u = \sigma^\infty \sqrt{2ar} \, \mathrm{Re} \, \{ \frac{1}{(S_1 - S_2)} [S_1 p_2 \psi_2^{1/2} - S_2 p_1 \psi_1^{1/2}] \}$$

$$+ \tau^\infty \sqrt{2ar} \, \mathrm{Re} \, \{ \frac{1}{(S_1 - S_2)} [p_2 \psi_2^{1/2} - p_1 \psi_1^{1/2}] \} \qquad (10)$$

$$v = \sigma^\infty \sqrt{2ar} \, \mathrm{Re} \, \{ \frac{1}{(S_1 - S_2)} [S_1 q_2 \psi_2 - S_2 q_1 \psi_1] \}$$

$$+ \tau^\infty \sqrt{2ar} \, \mathrm{Re} \{ \frac{1}{(S_1 - S_2)} [q_2 \psi_2^{1/2} - q_1 \psi_1^{1/2}] \}$$

where

$$\psi_1 = \cos \phi + S_1 \sin \phi \qquad\qquad \psi_2 = \cos \phi + S_2 \sin \phi$$

$$P_1 = A_{11}S_1^2 + A_{12} - A_{16}S_1 \qquad\qquad P_2 = A_{11}S_2^2 + A_{12} - A_{16}S_2$$

$$q_1 = \frac{A_{12}S_1^2 + A_{22} - A_{26}S_1}{S_1} \qquad\qquad q_2 = \frac{A_{12}S_2^2 + A_{22} - A_{26}S_2}{S_2}$$

As in the isotropic case, the crack tip stresses exhibit a singularity of $1/\sqrt{r}$. However, the magnitude of the stresses is not simply a function of the stress intensity factors. The quantities S_1 and S_2 also affect the magnitude of the stresses. This is an important difference between aniso-tropic and isotropic fracture. In anisotropic fracture, the magnitude of the crack tip stresses is a function of not only the applied load, specimen geometry and crack length, but also the material properties and the orienta-tion of the crack relative to the principal material direction. Application of this solution to the analysis of unidirectional composites with crack orientations other than parallel to the X-asis is presented in Gregory and Herakovich (9).

EXPERIMENTAL PROGRAM

In order to test the ability of the theoretical models to predict the direction of crack extension, tests were performed on center-cracked spe-cimens of 16 ply unidirectional AS4/3501-6 graphite/epoxy. Material proper-ties for AS4/3501-6 graphite/epoxy are given in Table 2.

Table 2 Lamina properties of AS4/3501-6 graphite-epoxy

E_1	= 21.6 MSI	(148.9 GPa)
E_2	= 1.96 MSI	(13.5 GPa)
G_{12}	= 0.83 MSI	(5.7 GPa)
ν_{12}	= 0.28	
X_T	= 282 KSI	(1.94 GPa)
X_c	= -282 KSI	(-1.94 GPa)
Y_T	= 10 KSI	(68.9 MPa)
Y_c	= -10 KSI	(-68.9 MPa)
S	= 14,2 KSI	(97.9 MPa)

The experimental investigation consisted of a series of 15° off-axis tensile tests with rigid end constraints. Coupons of various aspect ratios with

pre-machined cracks oriented perpendicular to the loading direction or perpendicular to the fibers were tested. The specimen configurations are illustrated in Fig. 4; two coupons were tested for each configuration.

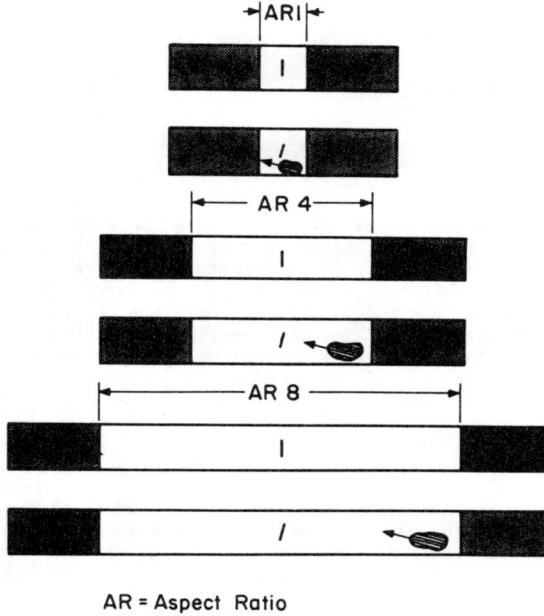

AR = Aspect Ratio

Fig. 4 Center cracked specimen configurations.

The test specimens were 1.00 inch (25.4 mm) wide and contained center
cracks of 0.20 inches (5.08 mm). The aspect ratios (gauge length to width)
of the specimens were 8, 4 and 1. Since off-axis specimens with rigid end
constraints experience increased shear loading as the aspect ratio is reduced (10), specimens of various aspect ratios were tested to vary the biaxial stress state in the region of the crack.

Strain gauge rosettes were attached to each specimen at a point far from
the crack. During each test, the direction of crack extension, the load
at crack initiation, and the load at failure were observed and measured.
To facilitate visual observation of crack growth, load was incrementally
applied at a crosshead speed of 40 microstrain per second.

CORRELATION OF THEORETICAL MODELS WITH EXPERIMENTAL RESULTS

In all the experiments performed, crack extension occurred parallel to
the fibers, with no observable fibre breakage. A broken specimen for each
combination of crack orientation and aspect ratio is shown in Fig. 5.

The anisotropic elasticity solution was used to model the experimental
procedures. To better approximate the far field stresses acting on the
crack, the Pagano and Halpin solution (10) for the state of stress in an

off-axis tensile coupon with rigid end constraints was incorporated. Though the Pagano and Halpin solution does not account for the presence of a crack, it does account for the high stress gradients and shear stress generated by the fixed ends. The far field stresses used as input for the anisotropic elasticity solution are the stresses generated by the Pagano and Halpin solution at a point corresponding to the crack tip location. The far field stresses, predicted direction of crack growth and the experimentally observed direction of crack extension are given in Table 3.

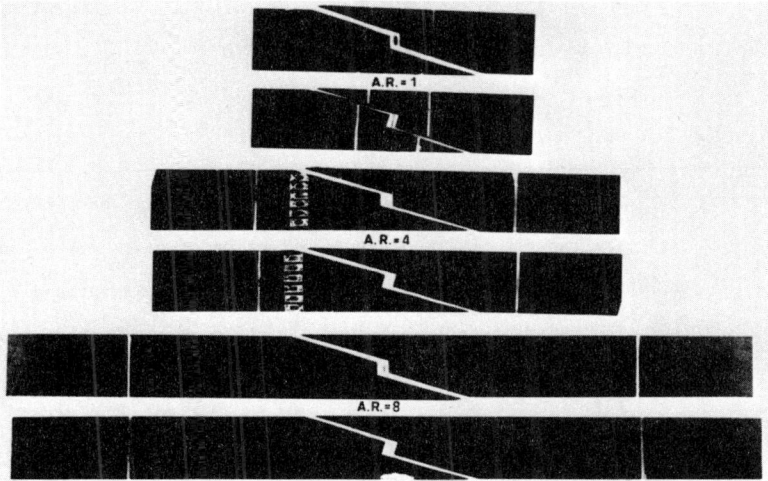

Fig. 5 Failed center cracked specimens.

From Table 3, it is apparent that only the normal stress ratio criterion predicts the correct direction of crack extension. The other crack extension direction criteria show no correlation with the experimental results. Distributions in the crack extension criteria as a function of ϕ, for test case A (crack perpendicular to the loading axis), with aspect ratio 1 are shown in Figs. 6-8. Distributions in the normal stress ratio as a function of , for aspect ratios 1, 4, and 8 of test case B, are shown in Figs. 9-11.

Analysis of Figs. 6, 9, 10 and 11 and Table 3 yields an interesting characteristic of the normal stress ratio criterion. The theoretical predictions of crack extension direction differ slightly from the experimentally observed values. There is, however, a strong peak in the distribution of the normal stress ratio as a function of ϕ in the actual direction of crack growth. This fact is very important. The normal stress ratio may not have the accuracy to predict correctly the direction of crack extension to within one degree. When observed graphically, however, the normal stress ratio represents the direction of crack extension exceptionally well.

Table 3 Comparison of theoretical and experimental results.

Aspect ratio	Test case	s_{yy} (KSI)	t_{xy} (KSI)	Predicted direction of crack extension		
				Normal stress ratio	Tensor Polynomial	Strain energy density
1	A	13.7	2.86	72°	347°	0°
4	A	10.2	1.01	73°	85°	315°
8	A	8.98	0.33	74°	87°	305°
1	B	13.7	-2.86	275°	10°	330°
4	B	10.2	-1.01	87°	110°	325°
8	B	8.98	-0.33	88°	103°	315°

Notes :
(i)   Specimens from test case A have center cracks perpendicular to loading axis.
(ii) The experimentally observed direction of crack extension for all specimens of test case A is 75°.
(iii) Specimens from test case B have center cracks perpendicular to the fiber direction.
(iv) The experimentally observed direction of crack extension for all specimens of test case B is 90°.
(v)   The far field loads correspond to the crack tip loading at 1 % strain.

The normal stress ratio criterion predicted the correct direction of crack extension for every test analyzed, except for test case B (crack perpendicular to the fibers) with aspect ratio 1. For this problem, the normal stress ratio correctly predicts crack extension parallel to the fibers, however the predicted direction of extension is 180° out of phase with the observed direction. Analysis of Figs. 9-11 reveals that for test case B, there are two peaks in the normal stress rato. The first peak, near $\phi = 90°$, predicts crack extension parallel to the fibers, toward the center of the coupon. The second peak, near $\phi = +90°$, implies crack extension parallel to the fibers, toward the free edge. For aspect ratios 4 and 8, the second peak is the maximum value. This is not true for aspect ratio 1, which has a maximum value at the first peak.

The discrepancy in the normal stress ratio criterion for test case B with aspect ratio 1 does not necessarily compromise the validity of the criterion. The stress gradients in a rigidly constrained tensile test of aspect ratio 1 are very high. One can also question the adequacy of the crack tip stress analysis; i.e., using Pagano and Halpin tensile coupon stresses as far field stresses in the anisotropic elasticity solution. The discrepancy is noted and further research is required.

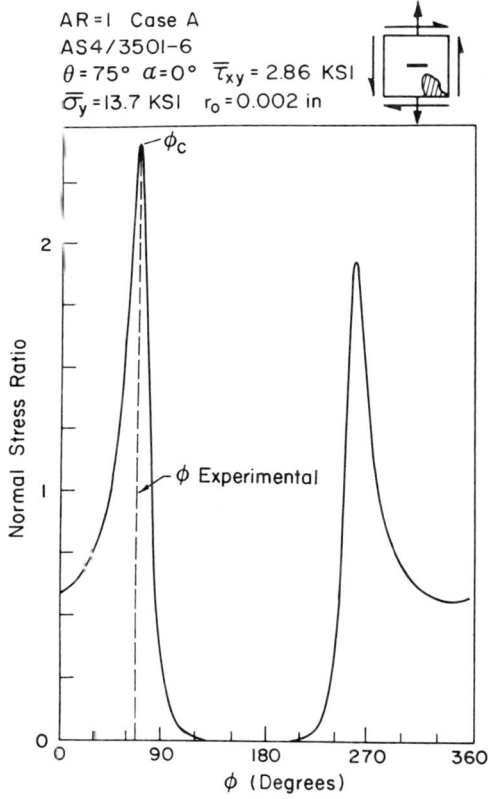

Fig. 6 Normal stress ratio near a crack tip, AR = 1, case A.

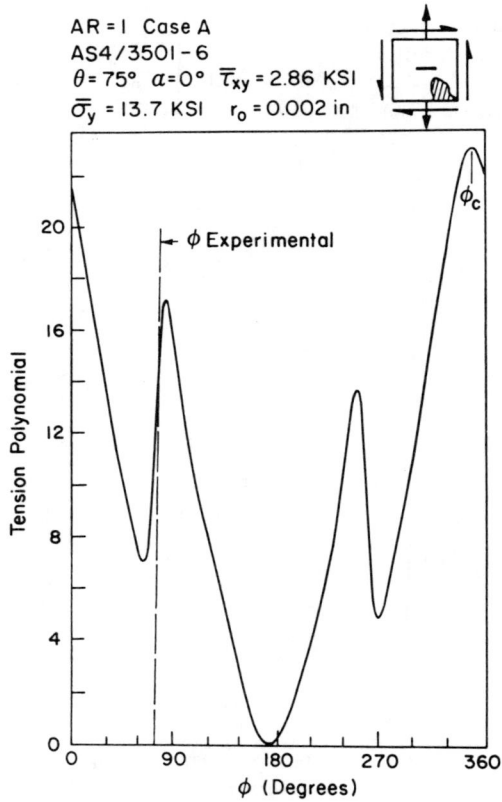

Fig. 7 Tensor polynomial near a crack tip, AR = 1, case A.

109

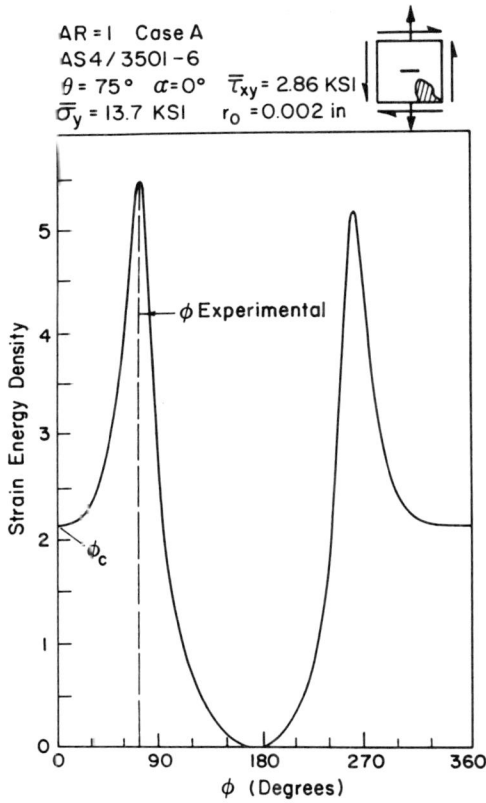

Fig. 8 Strain energy density near a crack tip, AR = 1, case A.

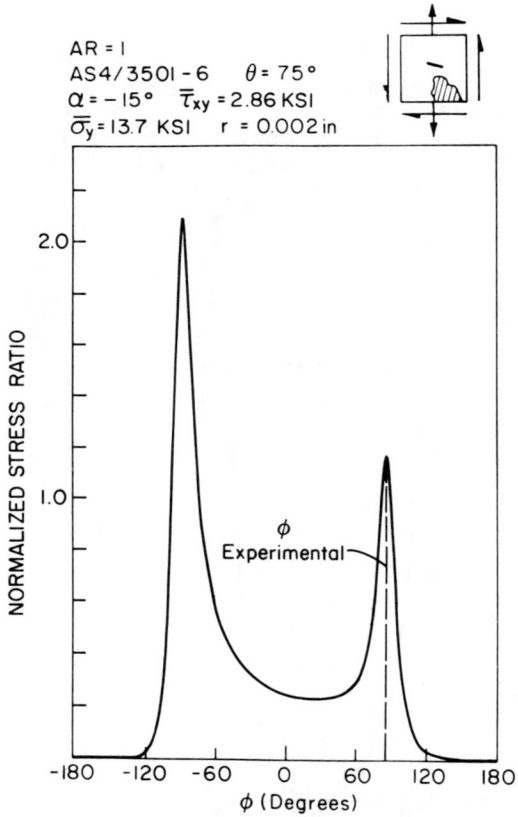

Fig. 9 Normal stress ratio near a crack tip, AR = 1, case B.

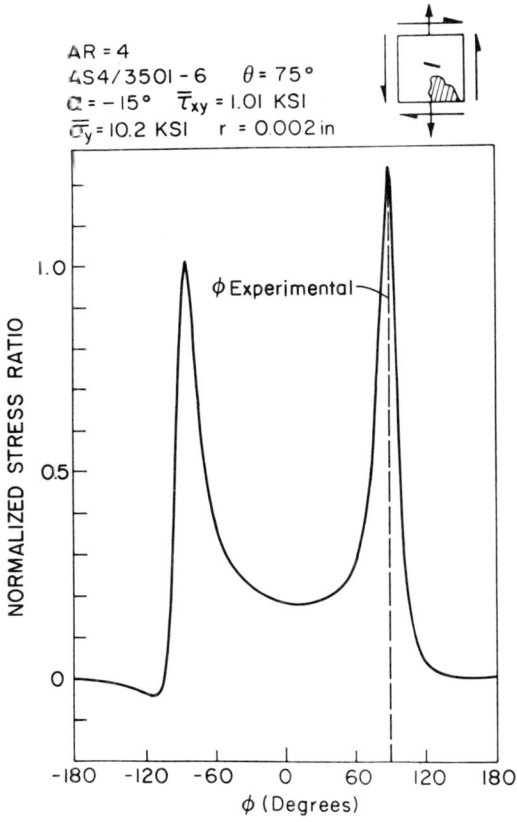

Fig. 10   Normal stress ratio near a crack tip, AR = 4, case B.

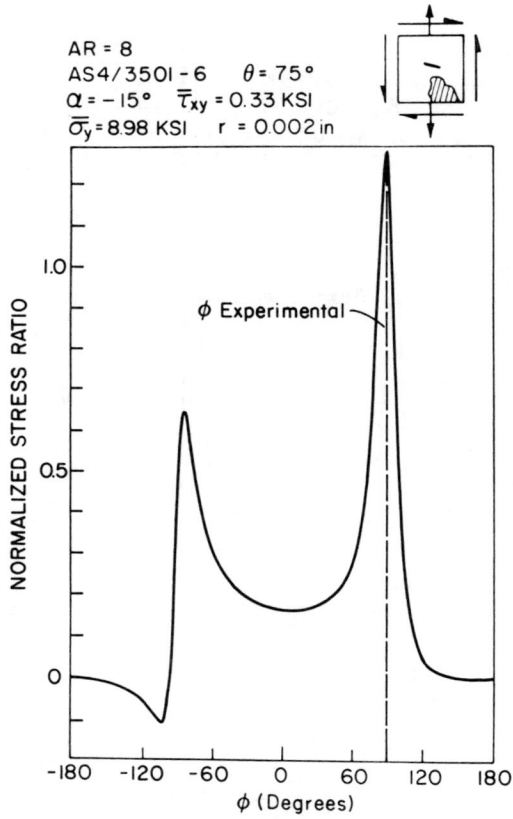

AR = 8
AS4/3501-6 $\theta = 75°$
$\alpha = -15°$ $\overline{\tau}_{xy} = 0.33$ KSI
$\overline{\sigma}_y = 8.98$ KSI   r = 0.002 in

ϕ Experimental

NORMALIZED STRESS RATIO

ϕ (Degrees)

Fig. 11 Normal stress ratio near a crack tip, AR = 8, case B.

CONCLUSIONS

This study was concerned with the development of a model to predict the direction of crack extension in unidirectional composite materials. An anisotropic elasticity solution, in conjunction with Pagano and Halpin's solution for stresses in a fixed end tensile test, was used to calculate the crack tip stress field in a center-cracked off-axis tensile coupon. Three crack extension direction criteria, the normal stress ratio criterion, the tensor polynomial criterion and the strain energy density criterion, were then incorporated into the model to predict the direction of crack extension.

Comparison of the predicted direction of crack extension with experimentally observed crack growth, indicates that only the normal stress ratio criterion consistently predicts the correct direction of crack extension.

ACKNOWLEDGEMENTS

This work was supported by Hercules Incorporated and the NASA Virginia Tech Composites Program, NAG-1-343. The authors are grateful for this support. The authors also want to acknowledge the many helpful discussions with professor C.W. Smith.

REFERENCES

1. Herakovich, C.I., Influence of Layer Thickness on the Strength of Angle-Ply Laminates.
   Journal of Composite Materials, vol. 16, pp. 216-227, 1982.

2. Buczek, M.B. and Herakovich, C.T., Direction of Crack Growth in Fibrous Composites.
   Mechanics of Composite Materials 1983, AMD-vol. 58, edited by G.J. Dvorak, American Society of Mechanical Engineers, New York, pp. 75-82, 1983.

3. Wu, E.M., Strength and Fracture of Composites.
   Composite Materials, vol. 5, edited by L.J. Broutman, Academic Press, New York, pp. 151-247, 1974.

4. Sih, G.C., Chen, E.M., Huang, S.L. and McGuillen, E.J., Material Characterization on the Fracture of Filament-Reinforced Composite.
   Journal of Composite Materials, vol.9, pp. 167-186, 1975.

5. Tsai, S.W. and Wu, E.M., A General Theory of Strength for Anisotropic Materials.
   Journal of Composite Materials, vol. 5, pp. 58-80, 1971.

6. Sih, G.C., A Special Theory of Crack Propagation.
   Method of Analysis and Solutions to Crack Problems, edited by G.C. Sih, Wolters-Noordhoff, pp. XXI-XLV, 1972.

7. Lekhnitskii, S.G., Theory of Elasticity of an Anisotropic Elastic Body.
   English translation by Brandstatton, Holden-Day Inc., San Francisco, 1963.

8. Wu, E.M., Fracture Mechanics of Anisotropic Plates.
    Composite Materials Workshop, edited by S.W. Tsai, J.C. Halpin and N.J.
    Pagano, Technomic Press, Stamford, CT, pp. 20-43, 1968.

9. Gregory, M.A. and Herakovich, C.T., Prediction of Crack Extension Direc-
    tion in Unidirectional Composites.
    CCMS-84-11, VPI-E-84-27, Virginia Polytechnic Institute and State Univer-
    sity, August 1984.

10. Pagano, N.J. and Halpin, J.C., Influence of End Constraints in the Tes-
    ting of Anisotropic Bodies.
    Journal of Composite Materials, vol. 2, pp. 18-31, 1968.

CONTINUUM DAMAGE APPROACH TO MECHANICAL BEHAVIOUR OF DAMAGED LAMINATE AND A MODELLING OF DAMAGE PARAMETER

K. KAMIMURA

Université de Compiègne, France.

ABSTRACT

Recent studies show that the prediction of realistic mechanical behaviour of composite materials needs to account of the existence of real damage growth. Our studies involved the establishment of global damage behaviour of composite laminates. The damage model is introduced in conjunction with the continuum damage approach in classical laminate plate theory. The good relation between experimental and theoretical results justifies our approach to the prediction of damaged behaviour.

INTRODUCTION

In spite of recent sustained effects for prediction of long-term mechanical behaviour of composite materials, this subject is not yet fully realised. The obstacle is mainly due to the variety of damages and their random apparitions. Recent studies show that the failure mode of fiber composite materials can be classified by (Fig. 1) :

- matrix cracking (transverse cracking)
- delamination
- fiber breaking.

In particular, the matrix cracking occurs by random manners in global fields. This type of damage is important because of :

- the reduction of dynamic mechanical properties such as the stiffness and the strength (1)

- the possibility of significant influence by environmental conditions such as a thermal cycling and a moisture absorption (2).

Thus the matrix cracking condition can be one of the key factors for the reliable design of composite materials and their structures. The investigations of this damage show that :

- the matrix cracking can reach a saturated level independent of loading histories (3)

- the loss of mechanical properties is closely related to the damage accumulation (4).

Our current study is intended to describe the damage growth mechanism and to predict the mechanical behaviour of damaged composite materials.

The laminate plate theory is applied to the continuum damage approach. This approach can include the various modes of damage under various loading conditions. Damaged elastic plies behaviours are defined by damaged stiffness matrix which is composed of three parameters : longitudinal, transversal and shear damage (D_1, D_2 and D_6). In this paper, the matrix cracking is simulated on the behaviour of damaged composite laminates.

ANALYSIS

Constitutive equations

The damage model is introduced in conjunction with the continuum damage approach (5) in classical "laminated plate theory" (6). The advantages of the continuum damage approach are :

- parametric description of mechanical behaviour of damaged materials
- modelling of real damage growth
- possibility for progressive improvement of damage model.

The continuum damage approach assumes that the stress state of materials is not constant after the damaging. The effective stress ($\hat{\sigma}$) on damaged materials is :

$$\hat{\sigma} = \sigma/(1-D) \tag{1}$$

in which the damage parameter D defines the damaged state of materials. For example, D is the variation of stressed area induced by damage :

$$D = (1 - \frac{S_{eff}}{S_o}) \tag{2}$$

S_o : non-damaged section ; S_{eff} : damaged section.

The difficulty arising from this expression is :

- the quantitative measurement of the surface change or of the volume change is not easy (7)
- the extension for multidimensional condition, for example, the damaged elastic relation produces the antisymmetric matrix compliance $S_{ij} \neq S_{ji}$ ($i \neq j$).

If we assume that the damaged materials behave elastically

$$\hat{\sigma} = E_o \varepsilon \tag{3}$$

E_o is Young modulus of non-damaged materials.
Substituting equation (1) by equation (3), the classical stress under damaged condition is :

$$\sigma = E_o(1-D) \tag{4}$$

$$= \hat{E} \varepsilon \tag{5}$$

$\hat{E}$ is the effective rigidity of damaged materials and by the equation (5).

$$D = (1 - \hat{E}/E_o) \tag{6}$$

Such a condition is available for only in unidimensional field and the equation (1) and (6) are no longer equivalent for multidimensional field.
Figure 2 shows the graphical presentation of the equation (6) under the uniaxial tension. If the material produces the highly oriented damage, the
Poisson's deformation of damaged materials (along the crack direction) is
not so much affected by the occurence of damage.

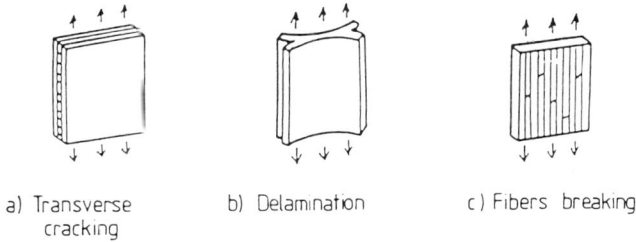

a) Transverse   b) Delamination   c) Fibers breaking
cracking

Fig. 1   Failure modes in fiber reinforced composites.

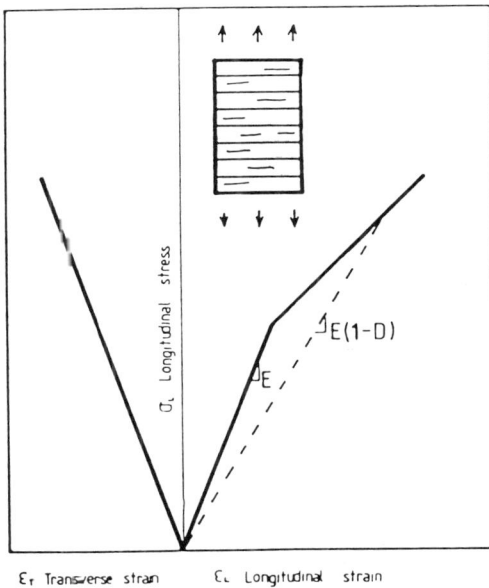

Fig. 2   Damaged response of unidirectional composite under transverse
tensile loading.

Figure 3 shows that the failure modes of unidirectional fiber composites laminae presents such types of damage. So we assume that the damage definition of equation (6) is available for in-plane behaviour of fiber reinforced orthotropic lamina , in which the three types of damage parameters exist, such as D_1, D_2 and D_6. The effective engineering constants of orthotropic laminae are :

$$\hat{E}_1 = E_1(1 - D_1)$$

$$\hat{E}_2 = E_2(1 - D_2) \tag{7}$$

$$\hat{G}_{12} = G_{12}(1 - D_6)$$

by these equations, we deduced the following damaged Poisson's ratio :

$$\hat{v}_{12} = \hat{v}_{12}(1 - D_1)$$

$$\hat{v}_{12} = \hat{v}_{21}(1 - D_2) \tag{8}$$

Taking account of the equations (7) and (8) in the stress-strain relation of orthotropic laminae, the damaged elastic behaviours of laminae are :

$$
\begin{Bmatrix} \varepsilon_1 \\ \varepsilon_2 \\ \gamma_{12} \end{Bmatrix}
=
\begin{vmatrix}
\hat{S}_{11} & \hat{S}_{12} & 0 \\
 & \hat{S}_{22} & 0 \\
\text{sym.} & & \hat{S}_{66}
\end{vmatrix}
\begin{Bmatrix} \sigma_1 \\ \sigma_2 \\ \tau_{12} \end{Bmatrix}
\tag{9}
$$

$\hat{S}_{ij}$ is effective compliance of damaged laminae ;

$$\hat{S}_{11} = \frac{1}{E_1(1 - D_1)}$$

$$\hat{S}_{12} = \hat{S}_{21} = -\frac{v_{21}}{E_1} = -\frac{v_{21}}{E_2}$$

$$\hat{S}_{22} = \frac{1}{E_2(1 - D_2)} \tag{10}$$

$$\hat{S}_{66} = \frac{1}{G_{12}(1 - D_6)}$$

The equations (9) and (10) show that the symmetry of compliance is assured by the equation (8). Some authors (8), (9) define the same types of relations, but the nature of damage parameter is not clearly defined. The non-damaged behaviour is determined by setting $D_1 = D_2 = D_6 = 0$.

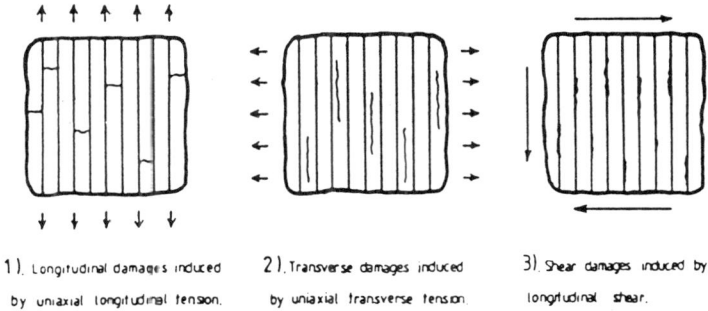

1). Longitudinal damages induced by uniaxial longitudinal tension.

2). Transverse damages induced by uniaxial transverse tension.

3). Shear damages induced by longitudinal shear.

Fig. 3   Damage modes of unidirectional composites under uniaxial loading
on special orthotropic axes.

The damage initiation of ply is determined by laminar failure criteria which
allow to distinguish the different failure modes such as those of Puck (10)
or of Hashin (11).
For the laminate, the damaged mechanical behaviour is easily extended by
using the classical laminate theory (6). We note that the damage accumulation of continued layers in the laminate, such as the matrix cracking of
transverse layers, does not appear under loading of an individual transverse
ply which breaks rather brittle than progressive. This is due to the non-
damaged longitudinal layer which sustains the accumulation of matrix cracking.
Consequently, the transverse damage parameter D_2 in the equation (10) is fic-
tive coefficient on the in-plane damaged behaviour of individual laminar.
Also, the experimental determination of D_2 must be done on the laminate.

Modelling of transverse damage parameter
    If we take account of the precedent considerations, the experimental de-
termination of damage parameter is possible, but the same experimental
works (12),(13) show that, for example, the matrix cracking accumulation is
significantly affected by the thickness of constituted layers.
This does not mean immediately the breakdown of the equation (9), but it
is necessary to check the real relation between the damage accumulation
and the damage parameter.
The matrix cracking of $(0/90)_s$ laminate under uniaxial tension is used in
order to check out the influence on constituted layers thickness on the
damage growth of laminate. The selection of this laminate is found on
following reasons :

- the strain capacity of non-damaged constituted layers (0°) is large to
  observe sufficiently the accumulation of matrix cracking in the (90°)
  layer

- the Poisson's ratio of non-damaged constituted layer (0°) is the smallest
  to avoid the influence of in-plane transverse deformation.

Under the uniaxial tension of this laminate, the shear damage parameter
(D_6) will be set to zero. Under the static loading condition, we assume
that (0°) layer breaks brittly which allows to set $D_1 = 0$. This is not
valid any more if the loading is dynamic. Thus, we can consider only the
accumulation of matrix cracking in (90°) layer and that can be converted

to the transversal damage parameter (D_2). The damage model needs to be as simple as possible to put into the damaged constitutive equations (eq. (9)), so the matrix cracking growth is to be analysed by the well-known unidimensional model of shear lag analogy (14). Figure 4 shows the damage element model in which the distribution of axial stress in damaged layer can be

$$\sigma_2 = \frac{\sigma \ E_2}{E_c} \ \{1 - \frac{\cosh \lambda \ (\frac{L}{2} - X)}{\cosh \lambda \ \frac{L}{2}} \} \tag{11}$$

in which

$$\sigma_c = \frac{{}_1 t_1 + {}_2 t_2}{t_1 + t_2}$$

$$E_c = \frac{E_1 t_1 + E_2 t_2}{t_1 + t_2}$$

$$\lambda^2 = \frac{E_1 E_2 t_1}{E_1 t_1 + E_2 t_2} \ \frac{2(G_1 G_2 / t_1 t_2)}{G_1 / t_1 + G_2 / t_2}$$

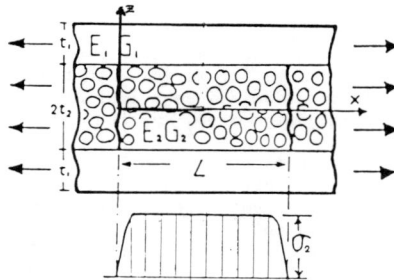

Fig. 4 Modelling of transverse cracking.

Assuming that the transversal layer cracks when the axial peak stress of transversal layer reaches to a critical value σ_{2f} at $X = L/2$, the first crack occurs when the laminate stress σ_{cf} is :

$$\sigma_{cf} = E_c \ \frac{{}_{2f}}{E_2} \tag{12}$$

Thre crack density is defined by :

$$d = \frac{1}{L} = 1 \ / \ \{\frac{2}{\lambda} \cosh^{-1} \ (\frac{1}{1 - \sigma_{cf}/\sigma_c}) \} \tag{13}$$

The average stress on the cracked element divided by the mean strain(ε_0) determines the effective modulus of cracked transversal layer :

$$E_{2eff} = \frac{2}{\varepsilon_0 L} \int_0^{L/2} \sigma_2 \, dx \qquad (14)$$

Finally, the transversal damage parameter is

$$D_2 = \frac{\sinh y}{y \cosh y} \qquad (15)$$

$$y = \frac{\lambda}{2d}$$

The prediction of D_2 needs only the constituted layers thickness (t_1 and t_2), the mechanical properties of constituted layers (E_1, G_1, E_2 and G_2) and the first ply failure which is thought to depend on the thickness. In this study, the theoretical prediction of σ_{2f} is not considered.

RESULTS

Evaluation of damage growth model
   The matrix cracking model is examined with the published experimental data (1). The table 1 shows the mechanical properties of layers which was used for this study. Figure 5 shows the variation of crack density versus laminate stress for glass/epoxy laminates $(0/90_3)_s$ and $(90_3/0)_s$.
A good correlation is obtained between theoretical and experimental data. The variation of crack density of transverse layer at free surface in $(90_3/0)_s$ laminate is theoretically identical with the one of middle layer in $(0/90_3)_s$. Figure 6 shows the variation of damage parameter (D_2) versus applied laminate stress (σ_c) of carbon/epoxy $(0/90)_s$ and $(0/90_2)_s$.

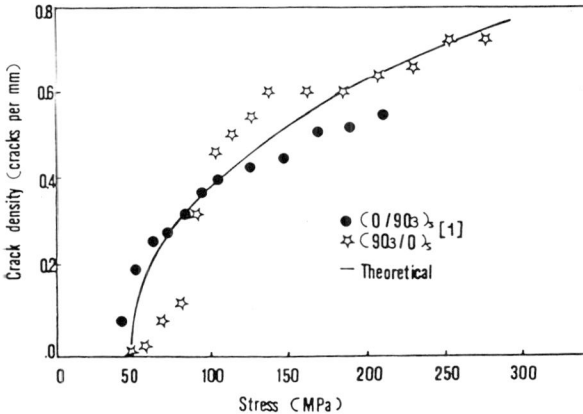

Fig. 5 Experimental and theoretical crack density vs. stress level for glass/epoxy laminate.

It is evident that the ply thickness effects   the damage parameter of two laminates. But, we find out that the difference is essentially induced by that of first ply failure stress level. For example, the lateral displacement of the D_2-σ_c curves on the same first ply failure level shows the minor difference of the variation of D_2. To check this tendency, the exact analytical modelling is needed, but the results of Nuismer and al. (15) shows also the same tendency with a different analytical approach. Thus, we assume that the D_2 curve is independent of ply thickness, and the ply thickness effects only on the first ply failure stress level.
This assumption is used for following analysis.

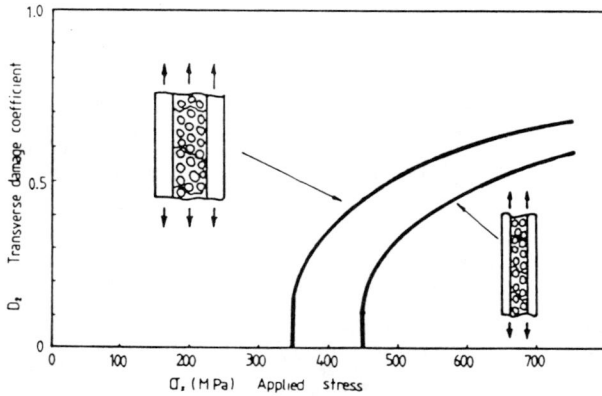

Fig. 6   Ply thickness effect on the variation of D_2.

Mechanical behaviour of damaged laminate.
Here, we assume that the mechanical behaviour of damaged laminate can be predicted by the strain depended function of damage parameter $D(\varepsilon)$ which can be determined by mechanical properties of laminar. The prediction of behaviour considered here is the $(0/90)_s$ laminate, so the damage parameters are $D_2 \neq 0$ and $D_1 = D_6 = 0$.
The engineering constants of $(0/90)_s$ laminate under the non-damaged condition are :

$$\nu_{xy} = \frac{A_{12}}{A_{22}}$$

$$E_x = \frac{A_{11}A_{22} - A_{12}^2}{h\,A_{22}}$$

$$E_y = \frac{A_{11}A_{22} - A_{12}^2}{h\,A_{11}}$$

(16)

in which the A_{ij} is the in-plane effective stiffness of laminate (6).

Under the damaged condition, the in-plane effective stiffness is reset by :

$$\hat{A}_{11} = A_{11}$$

$$\hat{A}_{12} = \{\frac{E_2 \, \nu_{12}}{1 - \nu_{12}\nu_{21}} + \frac{E_2 \, \nu_{12}(1-D_2)}{1 - \nu_{12}\nu_{21}(1-D_2)}\} \, \frac{h}{2} \qquad (17)$$

$$\hat{A}_{22} = \{\frac{E_2}{1 - \nu_{12}\nu_{21}} + \frac{E_1}{1 - \nu_{12}\nu_{21}(1-D_2)}\} \, \frac{h}{2}$$

Figure 7 shows the variation of stiffness versus crack density of glass/epoxy laminate $(0/90_3)_s$ and $(90_3/0)_s$.
The predicted result shows a good agreement with the experimental data (1).
The damaged behaviour of two laminates is essentially the same. Figure 8 shows the variation of crack density determined under tensile cycle loading of glass/epoxy $(0/90_3)_s$ laminate (1). The theoretical crack density was estimated by the stiffness variation of some specimen determined experimentally. This prediction is sequential and the dynamic parameter is not used here. The correlation is not good but the general tendency of the density variation is predicted. This result indicates the necessity to contain the dynamic damage growth parameter.
The complete characterization of non-damaged and damaged behaviour of carbon/epoxy $(0/90_2)_s$ under uniaxial tensile loading is examined. This laminate is composed in two thickness of transversal layers, but the first ply failure stress of these layers are not so much different. Consequently, we assume that the $\sigma_2 f$ of these layers are the same.

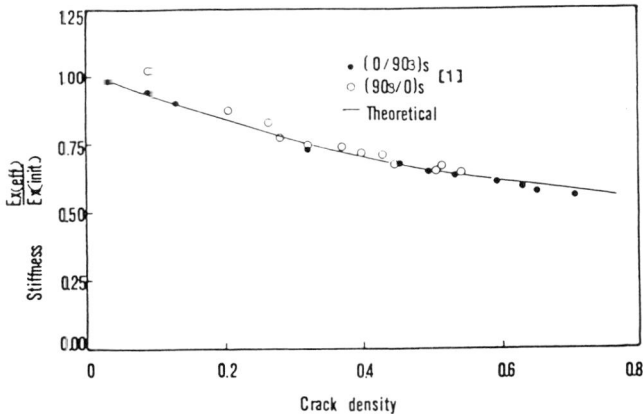

Fig. 7 Experimental and theoretical normalized stiffness versus crack density for glass/epoxy laminate.

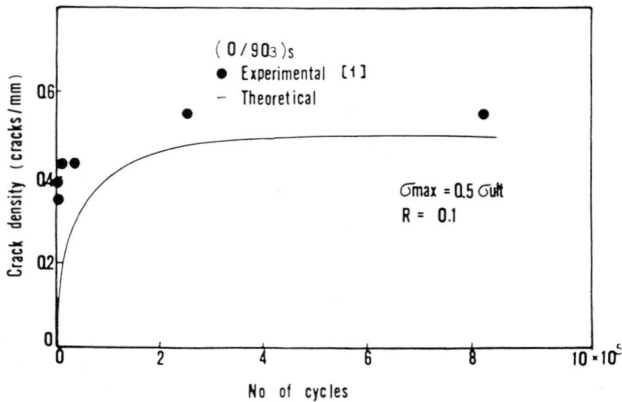

Fig. 8 Experimental and theoretical crack density versus cycles of loading
for glass/epoxy $(0/90_3)_s$ laminate.

Figure 9 shows the longitudinal and transversal stress-strain responses of
this laminate (16). The longitudinal response show the increasing of stiffness. This is contradictory to our prediction in which the damage produces
the decreasing of stiffness. So we conclude that the increasing of longitudinal stiffness is caused by strain which is induced by the hardening of
carbon fiber (17). The transversal response of this laminate shows clearly
the change of strain induced by damage. Note that the experimental determination of transversal strain of this laminate needs a great precaution
because of very small strains level (for example, the consideration of transversal sensibility of strain gauge).
The predicted transversal response correlates well with experimental determined response. The minor discrepancy between theoretical and experimental
response is observed. This may be caused by the approximate setting of
first ply failure stress level in two different thickness transversal layers.
We also note that in the case of transversal layer is more clearly appeared
on transversal response which agrees with early observations of Whitney and
al. (18).

CONCLUSION

The mechanical behaviour after damaging can be implied in the continuum
damage parameter of the classical laminate plate theory.
The damage parameter can be determined by experiments or by modelling the
damage element.
The anisotropic nature of damage must be taken account on the description
of damaged mechanical behaviour.
The theory presented here is a very flexible and physical meaning full
approach and that can be extended to many kinds of loading.

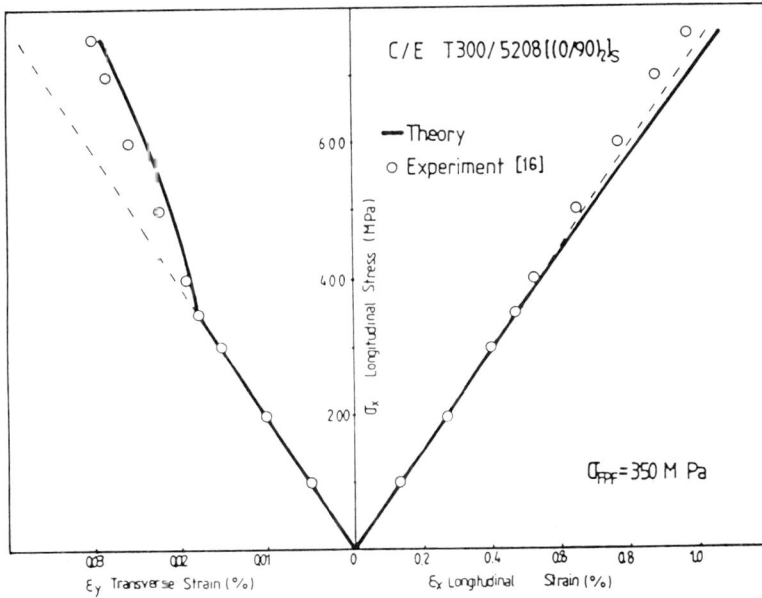

Fig. 9 Stress-strain curves for uniaxial tension of a $((0/90)_2)_s$ laminate.

REFERENCES

1. Highsmith, A.L. et.al., Stress-Reduction mechanisms in composite laminates. AST-STP-775, 1982, pp. 103-117.

2. Tompkins, S.S. et al., Effects of Thermal Cycling on Residual Mechanical Properties of C6000/MR-75 Graphite Polymide. 23rd SSDM Conf.,
   N.O., 1982, AIAA-82-0710.

3. Reifsnider, K.L. et al., Characteristic Damage States a New Approach
   to Representing Fatigue Damage in Composite Laminate. Materials, experimentation and design in Fatigue, Westbury House, Guildford, U.K., 1981.

4. Reifsnider, K.L. et al., Defect Property Relationships in Composite
   Materials, AFML-TR-76-81, Part 1, 1979.

5. Lemaitre, J. et al., Aspect Phénomenologique de la Rupture par Endommagement. J. Mécanique Appliquée, V. 2, no. 3, 1978.

6. Jones, R.M., Mechanics of Composite Materials. McGraw-Hill, New York,
   1975.

7. DUFAILLY, J., Modelisation mécanique et identification de l'endommagement plastique des métaux. Thèse de 3ème cycle, ENSET, 1981.

8. Nuismer, R.J. et al., Predicting the performance and failure of multi-
   directional polymeric matrix composite laminate. A combined micro-
   macro approach. Proc. 3rd Int. Conf. Composite Materials, Paris, 1980,
   pp. 436-452.

9. Poss, M., Endommagement et rupture des matériaux composite carbone-
   carbone. Thèse de 3ème cycle, ENSET, 1982.

10. Puck, A. et al., On Failure Mechanisms and Failure Criteria of Filament
    Wound Glass-Fiber/Resin Composites. Plastics and Polymer, Febr. 1969,
    pp. 33-43.

11. Hashin, Z., Failure Criteria for Unidirectional Fiber Composites.
    J. Appl. Mech., 1980, V.47, pp. 329-339.

12. Parvizi, A. et al., Constrained Cracking in Glass Fiber Reinforced
    Epoxy Cross-Ply Laminates". J. Materials Science, V.13, 1978, pp. 195.

13. Bader, M.G. et al., The Mechanisms of Initiation and Development of
    Damage in Multiaxial Fiber Reinforced Plastic Laminates. Proc. 3rd.
    Int. Symp. Mech. Behaviour of Materials, 1979, V.3, pp. 227.

14. Kamimura, K., Modelisation de la croissance d'endommagement appliquée
    aux matériaux composite en plaque stratifié. Internal report, June 83,
    Université de Technologie de Compiègne.

15. Nuismer, R.J. et al., The Role of Matrix Cracking in the Continuum Con-
    stitutive Behaviour of a Damaged Composite Ply. pp. 437-448.

16. Yoon, B.I., Endommagement dans un essai monotone d'un stratifié carbone/
    epoxy (0/90)$_s$. Internal report, 1983, Universite de Technologie de
    Compiègne, UTC-GPC-83-3.

17. Soulezelle, B. et al., Influence des contraintes internes sur le com-
    portement des composites. DRET, no. 79.34.343.00, 1981.

18. Whitney, J.M., The Relationship between Significant Damage and the
    Stress-Strain Response of Laminated Polymeric Matrix Composites.
    Composite Materials in Engineering Design, ed. R.B. Noton, 1972,
    ASM, pp. 441-447.

FAILURE OF GRP LAMINATES WITH REGARD TO ENVIRONMENTAL EFFECTS

W.S. CARSWELL

National Engineering Laboratory, Glasgow, Scotland.

INTRODUCTION

It is now generally recognized that fibre composite materials, i.e. non-metallic materials reinforced with glass or carbon fibres, exhibit fatigue behaviour in that under cyclic loading at levels significantly below that required to cause failure under static loading, failure will occur in a finite number of load cycles. Similarly in creep, rupture or failure occurs in a finite time at load levels well below that to produce failure in load increasing situation.

The deterioration in strength is attributed to the accumulation of damage which is manifested microstructurally be debonding, cracking, fibre failure, etc. These forms of damage in the fatigue or creep rupture condition are similar to that produced in failure under monotonic load increase, but the growth of such damage and the rate of accumulation will depend on the form of the test and the nature of the construction.

Under cyclic loading, composite materials exhibit hysteresis (Fig. 1) which can be used as a measure of the damage occuring within the material. The form of the hysteresis loop is controlled by the stress strain behaviour of the material and the anisotropic behaviour may be reflected in the shape of the loop. During the course of a fatigue test the hysteresis loop changes progressively by :

a)   rotating towards decreased stiffness, ref. (1),(2)
b)   drifting in the direction of applied mean stresses

implying a progressive increase in residual strain (Fig. 2).

These hysteresis effects and associated changes in behaviour are bulk effects occurring the whole strained volume of the test-piece. Failure, on the other hand, occurs over a small region of the material when the conditions of damage lead to a rapid deterioration of the material. Changes in hysteresis or other measurements of change in performance may be attributed to the spread of damage in directions or regions which do not contribute to the final failure. These changes may be considered as summations of damage over the whole volume and failure arises from interaction of such damage within a small volume.

The study of such effects will provide an indication of the manner of growth or accumulation of damage, this parameter being related to cracking and deterioration in properties.

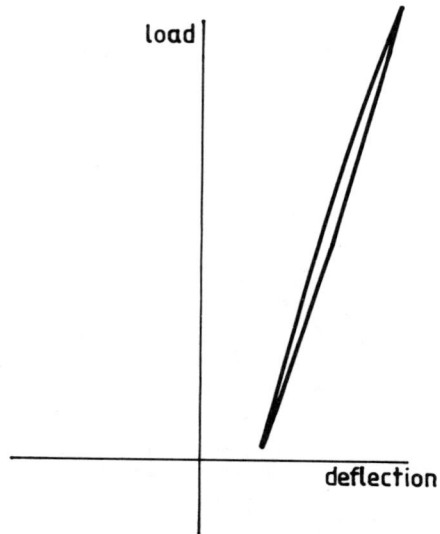

Fig. 1   Hystereses loop in tension only.

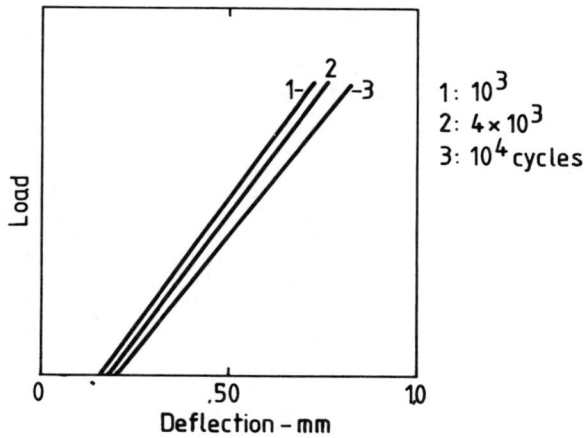

Fig. 2   Change in peak deflection with number of cycles.

MODEL OF FAILURE

The model of failure which is assumed here is the growth of two dimensional region of damage up to the point of catastrophic growth or failure. It is assumed that the area of damage A, is a function of two dimensions D and S of which D transverse to the loading direction is the dimension upon which failure depends, S is parallel to the loading direction.

It is assumed that generally in fatigue, ref. (3)

$$\frac{dD}{dN} = f(\sigma_c, D, S) \tag{1}$$

of simply

$$\frac{dD}{dN} = \sigma_c \frac{D}{S} \tag{2}$$

i.e. the rate of growth of the dimension D is a function of the stress concentration at tip of an elliptical damage region of dimensions D, and S.

The dimensions D and S are not independent and it has been assumed that the area A is a linear function of the number of cycles.
Then,

$$D\ S = K_1\ N \qquad \text{where } K_1 \text{ is a constant} \tag{3}$$

and

$$\frac{dD}{dN} = q_1\ \sigma_c \frac{D^2}{N} \qquad \text{where } q_1 \text{ is a constant} \tag{4}$$

On integration

$$\frac{1}{D_o} - \frac{1}{D} = q_1'\ \sigma_c\ \ell n\ N \tag{5}$$

where D_o is the damage extent at first load cycle.

At point of fracture some relation has to be assumed between D and σ_c the cyclic stress.
The simple relationship

$$\sigma_c\ D = K' = \sigma_u\ D_u \tag{6}$$

has been use here, where σ_u is stress at failure in tensile test ; D_u is the corresponding damage dimension ; K' a constant.

Similarly in a tensile test the rate of growth of damage with stress is assumed to be a function of the damage dimensions D and S, i.e.

$$\frac{dD}{dc} = q_2\ \sigma \frac{D}{S} \tag{7}$$

Also, the area of damage (DS) is assumed to be a linear function of σ, i.e.

$$A = (D\ S) = K_2\ \sigma \tag{8}$$

Thus from equations (7) and (8) integrating

$$\frac{1}{D_i} - \frac{1}{D} = q_2' \, (\sigma - \sigma_i) \tag{9}$$

where σ_i is stress at initiation of damage, D_i is damage dimension at point of initiation.

Combining expressions (5) and (9) eliminating D, the expression

$$\log N + p \, \frac{\sigma_i}{\sigma_u} = p \, \frac{\sigma_i}{\sigma_c} \tag{10}$$

or

$$N + a = \frac{b}{\sigma_c} \tag{11}$$

is obtained where $p = q_2'/q_1'$.

A similar expression can be derived for the time to failure related to the applied stress under creep rupture conditions.

RESULTS

The expression (11) has been used to evaluate the results of fatigue tests carried out on strips and creep rupture tests on tubes of chopped strand mat polyester laminates. The fatigue tests were carried out at a frequency of 15 Hz.

The fatigue test pieces were immersed in a liquid environment and a uni-axial cyclic load applied. In creep rupture the tubes were fitted with environment and pressurised with sealed ends.

The environments used were distilled water, 5 % dilute sulphuric acid and 5 % dilute sodium hydroxide (alkali).

The values of a and b obtained from these results using statistical method of analysis are given in Table 1.

The correlation factors ($\sim$ 90 %) obtained in the analysis was as good as that obtained using the more conventional method of presentation, viz

$$\log N + a^1 = b^1 \, \log \sigma_c$$

The use of greater powers of σ_c did not improve the correlation but rather clouded the physical significance of the parameters a and b.

From the creep rupture results the form of environment has a considerable effect on the values of both a and b ; the values decreasing with the severity of the environment. The same situation should apply in fatigue but is concealed by the scatter in results.

In every case in fatigue the values of an obtained were negative. This may be partially explained by proposing that fatigue damage is at least a two-stage process - the first being the slow accumulation of damage and the second the rapid growth of damage with macroscopic cracking.

Table 1 Value of parameters a en b in expression (11) for fatigue and
creep rupture tests.

Test	Environment	Temperature	a	b
Fatigue	water	RT	-3.06	78.0
		80°C	-4.47	22.4
	Acid	RT	-0.25	108.5
		80°C	-1.58	81.6
	Alkali	RT	-2.9	86.6
		80°C	-2.57	74.0
Creep rupture	Water	RT	6.4	294.0
		80°C	1.28	108.0
	Acid	RT	1.8	45.5
		80°C	0.96	28.36
	Alkali	RT	2.3	150.0
		80°C	2.4	122.0

ACKNOWLEDGEMENT

This paper is published with the permission of the Director of the National Engineering Laboratory, Department of Trade and Industry. Crown copyright is reserved.
The work report here was supported by the Mechanical and Electrical Engineering Requirements Board of the Department of Trade and Industry.

REFERENCES

1. Poursartip, A. Ashby, M.F. and Beaumont, P.W.R., "Damage accumulation in composites during fatigue". Fatigue and Creep of Composite Materials, Eds. H. Lilholt and R. Talreja, 3rd RISO Int. Symp., Denmark 1982.

2. Roylance, M.E., Houghton, W.W., Foley, G.E., Shuford, R.T. and Thomas, G.R., Characterization of cumulative damage in composites during service. AGARD Conference Proceedings no. 355, Neuilly-sur-Seine, France.

3. Carswell, W.S., "Fatigue in notched composites". Fatigue and Creep of Composite Materials, Eds. H. Lilholt and R. Talreja, 3rd RISO International Symposium, Denmark 1982.

ELASTIC RESPONSE AND DAMAGE ACCUMULATION OF COMPOSITE MATERIALS USED IN
PINNED-CONNECTIONS

E.D. KLANG and Th. DE JONG

Delft University of Technology, The Netherlands.

INTRODUCTION

The design of efficient joints for composite laminates used in structural
applications is an important engineering problem. Bolted and riveted joints
have received much attention in research work because they permit disassembly.
Knowledge of the stress distribution around the hole of a mechanical fastener
is of fundamental interest for predicting the strength and the failure mode
of the joint.

This work presents an elasticity solution and experimental results for a
finite width joint with one connector. Because of the two-dimensional
nature of the elasticity solution, bolt clamping forces and interlaminar
stresses were not considered. The solution was based on previous work,
ref. (1), in which fundamental aspects such as the role of friction, clearance and pin-flexibility were studied. Finite width effects were added
according to a method proposed in ref. (2).

A reason for the experimental work presented here was the need to under-
stand the failure modes and how damage accumulates in composite joints.
Another reason was the need for some check on the accuracy of the theoretical solution. The tests involved the accurate measurement of the elastic
response as well as the acoustic emissions of the specimens. The amplitude
distributions at different load levels were related to the results of X-ray
investigations of the damaged zones.

THEORETICAL

The elastic response of pin-loaded joints can be calculated using several
methods, however the accuracy of the solutions appears to depend strongly
on the use of correct boundary conditions. In general, there are two regions
of the composite panels where boundary conditions must be enforced. One
region is on the pin-hole interface where it is known that friction and
clearance play an important role. The second region includes all of the
far field boundary conditions. Infinite plate elasticity solutions do not
enforce these far field conditions in a realistic way. Therefore, these
solutions must be modified if good correlation between theory and experiment is expected. In this work, the solution of an infinite plate with
an unloaded hole was superposed with the pin-loaded infinite plate. The
superposition with finite geometry plates is illustrated in fig. 1.

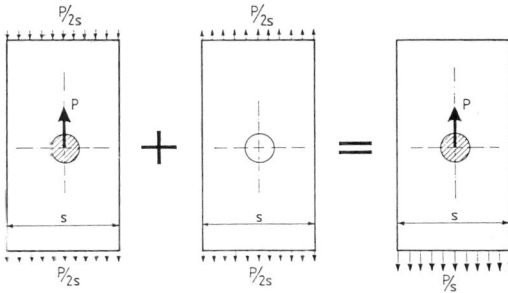

Fig. 1 The approximation for a finite geometry joint.

When infinite plates are used, the solution becomes an approximation. The essential feature of this superposition is that the far field load is reacted at only one end of the specimen rather than at both ends. Of course this is the situation for most experimental studies and is therefore quite important. The boundary conditions at the pin/hole interface were fulfilled using a collocation technique after the two infinite plate solutions were superposed. The regions over which the interface boundary conditions acted were not known a priori so iteration was required to find the correct solution.

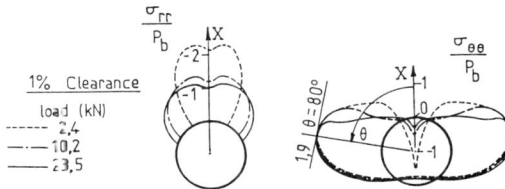

Fig. 2 The stresses at the hole edge for different load levels.

Figure 2 shows some results of this procedure for a $[0/\pm45/0]_{2S}$ graphite epoxy laminate. In this figure the radial and hoop stresses at the hole edge are shown for a joint with 1 % clearance. The stresses are normalized with the classical bearing stress ($p_b = P/Dt$).

EXPERIMENTAL

The specimens were configured so that bearing failures would occur. The nominal pin diameter was chosen to be 12 mm. Three graphite epoxy panels were constructed with stacking sequences of $[0/\pm45/0]_{2S}$, $[\pm45/0_2]_{2S}$ and $[\pm45/0]_{3S}$. The elastic response of the first two panels was similar but the failure loads were different. The third panel was chosen as a material with entirely different characteristics. A typical plot of load versus displacement for a specimen with 1 % clearance is shown in fig. 3.

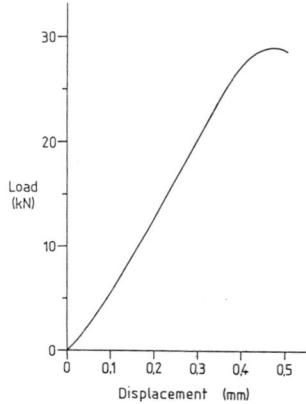

Fig. 3 Typical load-displacement curve.

The nonlinear region below 10 kN is due to the clearance between the pin and plate. Between 10 and 25 kN the response is nearly linear which indicates that the contact area between the pin and plate has reached an asymptotic value. Above 25 kN the specimen begins to fail and the stiffness falls off accordingly.

The damage accumulation at the higher loads was analyzed using acoustic emission and X-ray technique. The nonlinear elastic response, however, was compared with theoretical predictions. This comparison will be discussed first.

The modified elasticity solution showed that the clearance between the pin and hole has a measurable effect on the relation between pin load and pin displacement. Therefore this parameter was varied in the experimental program. Three values of clearance were chosen :

1) the zero clearance or push fit case where the hole diameter is the same as the pin diameter

2) one percent clearance, where the hole diameter is 1 % larger than the pin diameter

3) two percent clearance, where the hole diameter is 2 % larger than the pin diameter.

The elastic responses were recorded and some of these results appear in fig. 4, where a comparison of theory and experiment is provided.

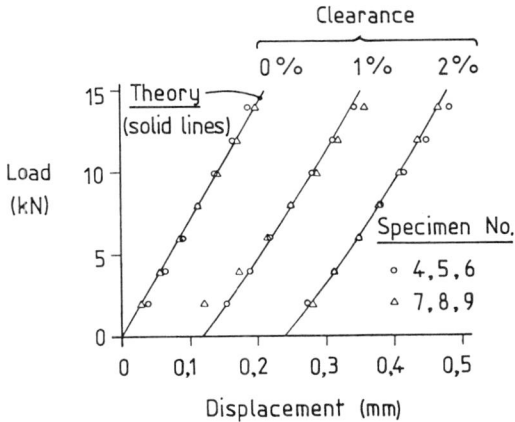

Fig. 4 Elastic response for different clearance values.

It is evident that the elasticity theory predicts the elastic response of these specimens rather well. The nonlinear responses for the clearance fit cases are also shown in this figure.

The acoustic emissions made by the specimens during the loading were recorded. The amplitude of each acoustic event is a strong indication of the damage accumulation and was therefore stored for plotting.

Figure 5 shows the amplitude distribution at two load levels. The group of relatively low amplitude events represents noise caused by a combination of the pin sliding and micro damage (wear-in). The ultimate failure however is characterized by relatively high amplitude events. Further nondestructive evaluations were carried out using TBE enhanced radiography.
Figure 6 shows an X-ray of a specimen at 85 % of its failure load.
This corresponds to the acoustic emission amplitude distribution of fig. 5 at a load level of 25 kN. It is apparent that no visable damage has occurred. As the load was incrased, the stiffness reduced (fig. 3) and the number of high amplitude events increased (fig. 5). The resulting damage state at the maximum load (29.3 kN) is also shown in fig. 6. The X-ray shows typical bearing damage which occurs just prior to large scale delamination after the peak load is passed.

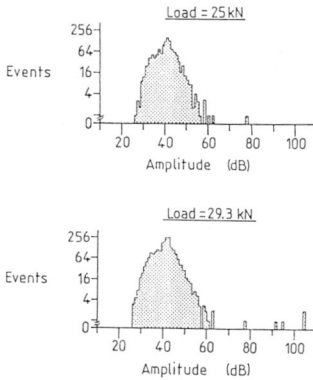

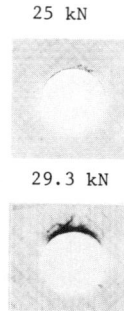

Fig. 5   Acoustic emission distributions   Fig. 6   X-ray results.

Each specimen was loaded to failure and the maximum loads were recorded. The loads were then used to calculate the classical bearing strengths which appear in table 1. The strength values are seen to decrease with increasing values of clearance. This indicates that the bearing strength is a function of geometry and not a material constant as is often assumed.

The classical bearing strength is defined as the failure load divided by the projected area of the hole. This definition is acceptable for cases where the contact region is subtended by a contact angle of 180°. However when clearance is present, the contact angle can be less than 180° and the projected contact area is no longer the same as the projected hole area. To correct this, the classical bearing strength can be modified by dividing it by the sine of the half-contact angle. The resulting modified bearing strength reflects the change in contact area for clearance fit joints and therefore may be a better choice as a material parameter. The calculated half-contact angles at failure and the modified bearing strength for each configuration are also listed in table 1.

It is interesting to note that the modified bearing strength for the $[\pm45,0_2]_{2S}$ laminate is much grater than that of the $[0,\pm45/0]_{2S}$ laminate. This is because the forty-five degree plies prevent out-of-plane buckling much more effectively than the zero degree plies. This effect may disappear when lateral constraints are added to the joint (i.e. a bolted joint).

Table 1 Bearing strength values for various configurations.

Laminate	Clearance	Half-Contact Angle β (deg.)	Classical Bearing Strength $\overline{P}_b$ (MPa)	Modified Bearing Strength $\overline{P}_b^*$ (MPa)
$[\pm45/0]_{3s}$	0%	92,7	458	458
	1%	74,5	440	457
	2%	61,0	414	473
$[\pm45/0_2]_{2s}$	0%	92,3	461	462
	1%	74,0	445	462
	2%	59,5	414	482
$[0/\pm45/0]_{2s}$	0%	92,3	413	413
	1%	70,0	381	405
	2%	58,0	357	420

$$\overline{P}_b = \frac{\text{failure load}}{D \cdot t}$$

$$\overline{P}_b^* = \frac{\text{failure load}}{(D \cdot \sin\beta)t} = \frac{\overline{P}_b}{\sin\beta}$$

CONCLUSIONS

From the research presented here, the following conclusions can be drawn :

1) Elasticity solutions with the correct boundary conditions predict the nonlinear elastic response of pin-loaded joints rather well.

2) The normal load acting on the hole of the plate differs substantially from the cosinusoidal distribution often assumed for finite element solutions.

3) Classical bearing strength values differ for various amounts of clearance and therefore they should be reported relative to the amount of clearance.

4) The modified bearing strength which accounts for clearance may be more suitable for use as a material property.

5) Failure of a pin-loaded composite joint is characterized by relatively high amplitude acoustic emissions which begin to occur at about 85 % of the ultimate load.

REFERENCES

1. Klang, E.C., The Stress Distribution in Pin-Loaded Orthotropic Plates.
   Ph.D. Dissertation, VPI & SU, Blacksburg, Virginia, August 1983.

2. DE JONG, Th., Stresses Around Pin-Loaded Holes in Elastically Orthotropic
   or Isotropic Plates.
   Journal of Composite Materials, vol. 11, July 1977.

BIAXIAL FATIGUE OF GLASS FIBRE REINFORCED POLYESTER RESIN SHEETS

J.C. RADON

Imperial College, London, United Kingdom.

ABSTRACT

Understanding the fracture phenomenon of a glass fibre reinforced polyester requires an accurate determination of the fatigue crack growth behaviour where the crack propagation rate governs the failure of the composite. Direct measurement of the crack length under fatigue loading is very difficult because of

a) the physical structure of the material, as a whole ; the gel coat interaction with the laminate and, microscopically, the glass fibre-resin matrix interaction ;

b) a substantial damage zone existing at the crack tip from which it is not clear whether the array of fibres behind the crack tip have broken or pulled out or both.

Fatigue crack growth tests were carried out on chopped strand mat polyester resin under biaxial tensile stress. Centre-notch specimens were used to measure the effects of the load biaxiality factor, B, defined as the ratio of the load acting normally to the crack line and that acting parallel to it. A compliance calibration technique was used to measure the crack growth so as to overcome difficulties with visual measurement when using conventional stress intensity factors in order to present the fatigue results on a fracture mechanics basis.
An adaptation of the compliance technique was also used to extend the applicability of the stress intensity factor concept to planar composite materials by proposing a K equation that takes into account the effects of reinforcement geometry and biaxial stress. This allows a safe life crack growth criterion to be used more effectively to assess damage tolerance taking account to a number of detrimental intrinsic effects due to the nature of the composite reinforcement.

The compliance technique was found to be reliable as long as the crack path direction was linear for biaxiality ratios between 0 and 1. However for biaxility ratios greater than 1 the crack path direction becomes unstable and S-shaped cracks may be generated by the fracture propagation. The moiré strain method was used to measure crack length in the two-dimensional plane ; it can be applied to running crack length measurement as well as to the measurement of crack displacement.

The Paris power relationship was found to be applicable to the results, the analysis of which shows that non-singular stresses do affect the behaviour of a crack subjected to plane stress cyclic loading.

Biaxial stresses were found to produce a shift in the fatigue crack propagation rates, notably a slight decrease in the Paris exponent with increase
in the load biaxility factor, B. Analysis of the fatigue behaviour indicates
the failure mechanisms are influenced by the reinforcement geometry of the
composite material, together with the Poisson's ratio of the material under
biaxial stress. It would seem that the fatigue behaviour is governed by
the fibre-resin interface at low stress levels, while at higher stresses the
Poisson's ratio of the composite determines the biaxial influence.

INTRODUCTION

Real materials do not exhibit linear elastic behaviour near the tip of a
flaw. A plastic zone develops at the crack tip to accomodate the crack tip
opening. This form of yielding at the crack tip, for example plastic flow
in metals, has the function of partially relaxing the high local stresses
and partially absorbing the fracture energy. The glass reinforced plastic
material investigated in the present work showed a small capacity for plastic
flow, and the yielding at the crack tip was observed to take the form of a
damage zone. This damage zone is a region of crack growth extending from
the tips of a precut centre notch and consisting of sub-critical secondary
cracks orientated along the interface between glass fibres and matrix all
around the main crack. This makes the direct measurement of crack length
very difficult because it is not always clear whether the aligned fibres
within the damage zone have either broken or pulled out, or where the crack
tip can be located within the thickness of the specimen. Thus substantial
variations in the physical crack length may be recorded.

Two methods were used for making crack measurement ; one, compliance ;
and the other a moiré fringe technique to circumvent difficulties observed
in the direct measurement of crack length in composite materials. A compliance technique has been developed to determine an equivalent crack length
which includes a plastic zone correction, using expressions derived from
fracture mechanics. The moiré fringe method provides a precise location of
the crack tip so that an accurate crack length measurement can be made.
No plastic zone correction was made for the moiré technique as plastic flow
ahead of the crack tip would have been observed as distortion in the fringe
pattern.

EXPERIMENTAL PROCEDURE

The cruciform specimens of 3.5 mm nominal thickness (1), were individually
moulded, the polyester laminates being impregnated with a powder bound E-
glass fibre in a chopped strand mat reinforcement configuration. A horizon-
tal testing machine of an electro-hydraulic type (2), cycling from zero to
a maximum load amplitude of 30 kN at a cyclic frequency range of 0.1 to 1 Hz,
was used to perform the fatigue tests.

Fatigue tests were carried out at a constant load range ΔP and biaxiality
factor B until fracture in two different environments, air at 20°C and 50 %
relative humidity, and 5 % sulphuric acid solution at 20°C, for B = 0 and 1
in each environment. The stress ratio equivalent to K_{min}/K_{max} and the constant load range rate dP/dt was maintained at 0.1 and 5 kN/sec. The cyclic
frequency f was related to the load range rate by f = 2.5/dP Hz.
For the acidic environment fatigue tests a chamber was constructed using

a rubber 'O' ring and polyethylene terephthalate film sealed together with
silicone rubber to both sides of the specimen. The compliance method was
used to determine the effective crack length, 2a, in relation to the biaxial
CN specimen compliance. Calibration curves for biaxility ratios B = 0, 1 and
2 are shown in Fig. 1, where the compliance, ϕ, is defined as the result
of crack opening displacement multiplied by specimen thickness and divided
by a corresponding load increment. The crack length was calculated and
plotted as a function of the number of cycles, N, from which the crack growth
rate was determined by graphical differentiation.

The plastic zone at the crack tip was originally developed by Irwin for
conditions of plane stress having a radius (a length) $r = K_I^2/2\pi\sigma_{ys}^2$.
The stress, $\sigma_{ys} = 52.7$ MPa, is the off-set stress obtained from the stress-
strain curve of the composite material, and corresponds to the stress at
which the resin cracking becomes a maximum. The modified half crack length
$a_m = a + r$ was plotted against the number of cycles, N, and subsequently
used in further calculations of da/dN and ΔK. An example of the plot of
the crack length versus cycles is shown in Fig. 2

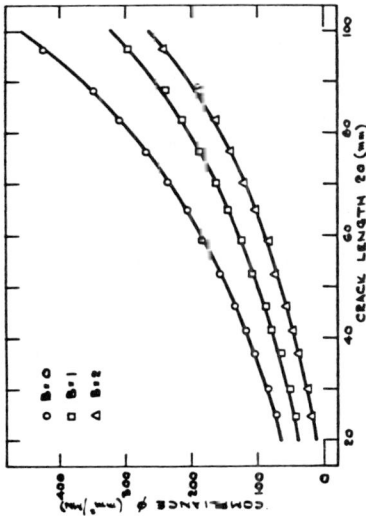

Fig. 1   Compliance calibration
curves

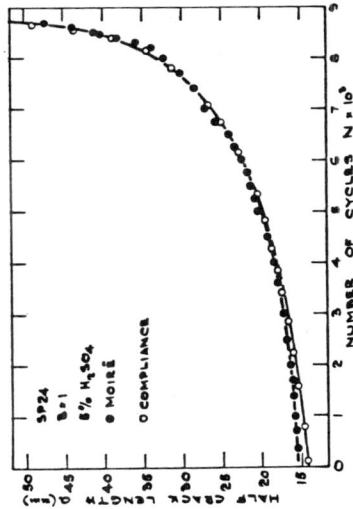

Fig. 2   Crack growth for B = 1 in
acid

For the moiré fringe method orthogonal gratings of 500 lines per inch were used, in the form of a film deposited on an acetate base. The gratings were cut to an appropriate size and then attached to the specimen. A second grating of the same pitch was superimposed on the fixed grating producing an interference pattern due to the misalignment of the gratings.

During a fatigue test the development of a crack disrupts the fringe patterns causing local kinks because displacement is not single valued at the crack tip. Location of the crack tip using moiré fringes and the grid attached to the specimen surface below the crack line, Fig. 3, was judged to be accurate to ± 0.1 mm or better. Fracture mechanics concepts were applied in analysing the crack propagation data obtained from the fatigue tests in the usual way (3).

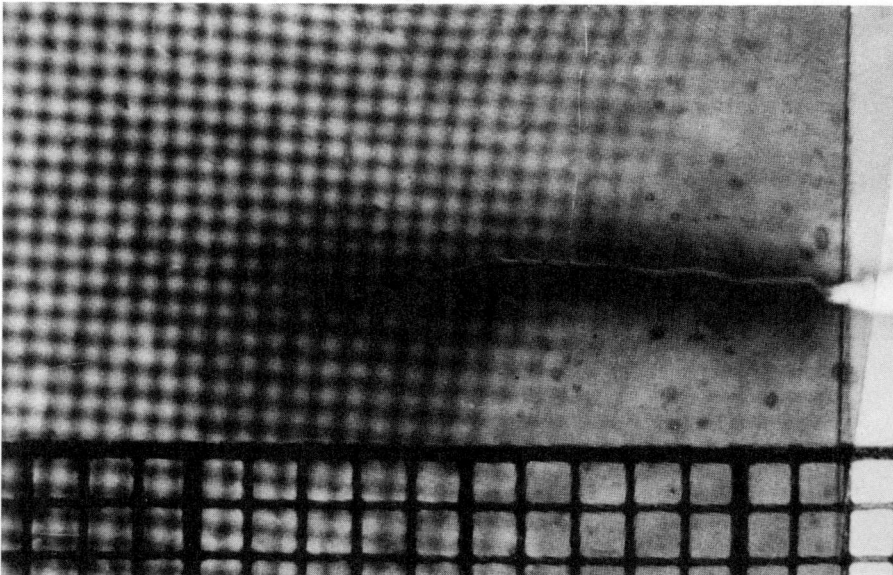

Fig. 3 Moiré fringe pattern for B = 1 in acid. Millimetre scale.

RESULTS AND DISCUSSION

In these constant load fatigue tests a cyclic plastic zone was created ahead of the crack tip. The boundary of this zone is usually expressed as a function of the angle θ and the length r_c, which is the radial distance from the crack tip,

$$r_c^\theta = \alpha^\theta \left(\frac{K_{max}}{\sigma_{cy}}\right)^2 \tag{1}$$

where σ_{cy} is the cyclic yield stress, which depends in particular on the environment and strain rate. As the yield stress usually increases with strain rate, the size of the plastic zone will decrease accordingly. Theoretically, the cyclic plastic zone will be a quarter of the monotonic plastic zone, as discussed in (4). Because the composite investigated here did not exhibit a sudden yield point, the off-set yield stress was used for the calculation of the corrected crack length.

The plastically deformed material, Fig. 4, was surrounded by the elastic field, where the stress and strain are proportional to $1/\sqrt{r}$. It can be divided into three zones :

(1) a monotonic plastic zone, experiencing low strain cycles of ε_p less than 10^{-3},

(2) a cyclic plastic zone, nearer to the fatigue crack tip, subjected to a higher strain of 10^{-1},

(3) an intensely deformed process zone, subjected to a very high strain of the order of 10^{0} at the tip of the crack, where the fracture took place. Fig. 4 shows the crack directly extending from the crack tip opening CTOD. The strain, proportional to $1/r$, is accommodated in local slip bands. The blunting of the crack tip, proportional to $K^2/\sigma_{cy} E$, leads firstly to the formation of local intense slip bands and secondly to the extension of the crack at each consecutive cycle. At the same time secondary cracks form parallel with the main crack, but mostly in the surface layer. This discontinuous movement of the crack tip at each cycle corresponds to the formation of striations as observed on the fracture surface in the region II of the crack propagation curve. The one-to-one correspondence with the fatigue striations does not take place in the region I. A gradual transition between these two regions was observed at the K_{max} values of 2.8 MPa m$^{1/2}$ for the biaxiality B equal to 0 and less than 2.2 for B = 1.

The fatigue crack growth in region II is usually expressed in the form of a da/dN versus ΔK curve using the Paris relationship (5),

$$\frac{da}{dN} = C \ (\Delta K)^m \qquad\qquad (2)$$

The constants C and m are determined from constant crack growth data of specimens cycled at a constant load range in steady state fatigue growth. The fatigue cycle was here described by $K = (K_{max} - K_{min})$; K_{max} and K_{min} represented the opening mode stress intensity factors calculated from the maximum and minimum stresses measured during the fatigue cycle for an appropriate stress ratio R. Because of the experimental limitations (2) it was advisable not to use a value of K_{min} below 10 % of K_{max}. It is interesting to note that for this GRP material the values of the exponent m were much higher than expected. In all the tests performed in air the value of m was 15 for B = 0 and decreased only slightly with increasing biaxiality (7). In the acidic environment this value changed to 5. It is well known from the literature that the respective values for metals are of the order of 2 to 4 and for polymers 3 to 8 (6).

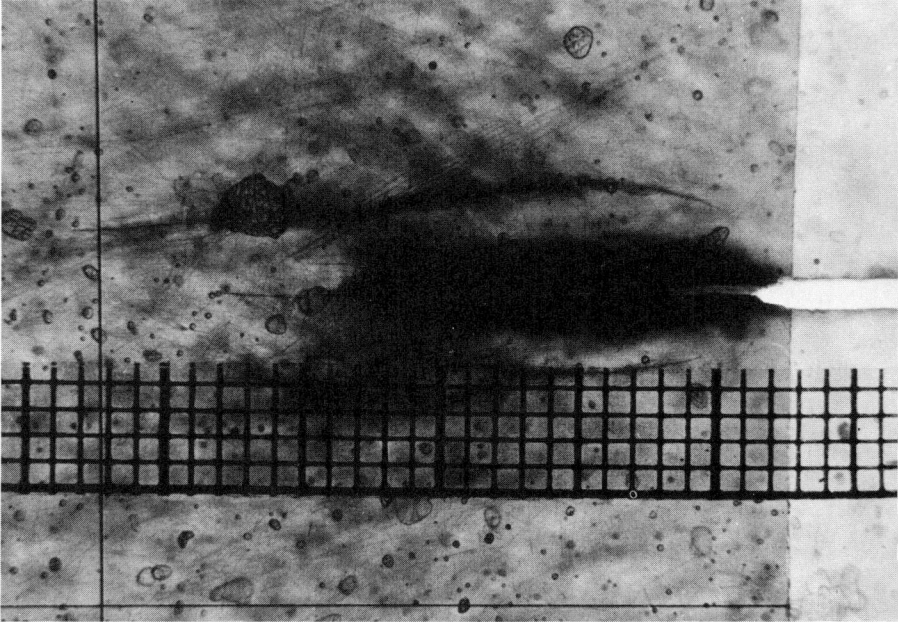

Fig. 4 Damage zone.

The environmental fatigue crack growth results under biaxial stress B equal to 0 and 1 are shown in Figs. 5 and 6, and analysis of the data shows a unique relationship existing between da/dN and ΔK for the two tests in air and in the acidic environment. The results also show fairly good agreement between moiré and compliance data for the conditions tested.

The analysis of the fatigue data has been discussed elsewhere in terms of environmental (7) and biaxiality (8) effects. The appearance of the fracture surfaces of the specimens from the dilute acid and air tests were found to be different ; those tested in air displayed significant fibre pull-out over the whole surface whereas the surface of the slow crack growth region of specimens fatigued in the acid environment showed little sign of fibre pull-out, giving a glassy appearance to the fracture surface. This lack of fibre pull-out is attributed to chemical attack of the fibres by acid, in conjunction with the mechanically applied cyclic stress, causing the fibres to fail along their length and effectively eliminating their reinforcing properties.

The damage zone in the context of this paper represents the crack and the debonded fibres around it, the consequence of environmental fatigue is the reduction in volume of that damage zone. The energy that would have been dissipated in the fibre pull-out mechanism was channelled into propagating the crack even further. The difference in crack growth data for acid and air tests is shown dramatically in Figs. 5 and 6, for comparative crack growth rates there is a reduction by a factor of two in the applied stress cycle, for the acid tests with a corresponding reduction in the volume of the damage zone.

The trajectories of fatigue cracks, particularly those in the region II, were reasonably straight. However, they showed a large amount of bifurcation and debris, the volume of which increased with the decreasing stress intensities. The degree of debonding appeared to be larger in air for all types of dynamic fractures, followed by an extensive pull-out of fibres. On the other hand, in the acidic environment the amount of debonding was much smaller and this was due to a large reduction in fibre strength. The crack initiated on the surface of the fibre immediately after the protective layer of PVA was leached away. Subsequently the crack propagated in the transverse direction of the fibre under the action of pure corrosive fatigue. With increasing ΔK the crack growth rate increased very slowly and it is expected that the growth rate in the region III would not differ greatly from that in air. A similar relative decrease in the rate was observed with increasing biaxiality B, Figs. 5 and 6. However, tests at much lower ΔK values are needed in order to evaluate the influence of environment on the fatigue behaviour of GRP. In the parallel fatigue tests on a low-alloy steel in NaCl solution (9) the crack growth rate decreased below that observed in a laboratory air environment at very low values of ΔK, thus showing that other factors, apart from the environmental attack may play an important role under cyclic loading.

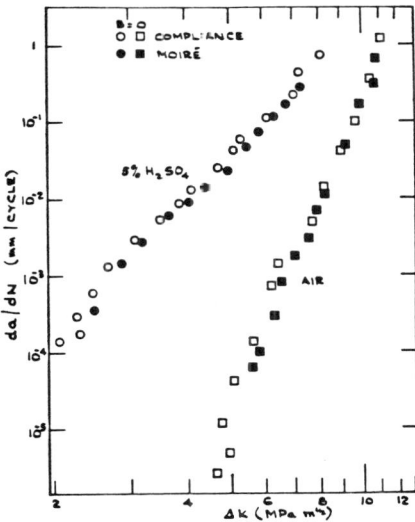

Fig. 5 da/dN vs. ΔK for B = 0.   Fig. 6 da/dN vs. ΔK for B = 1.

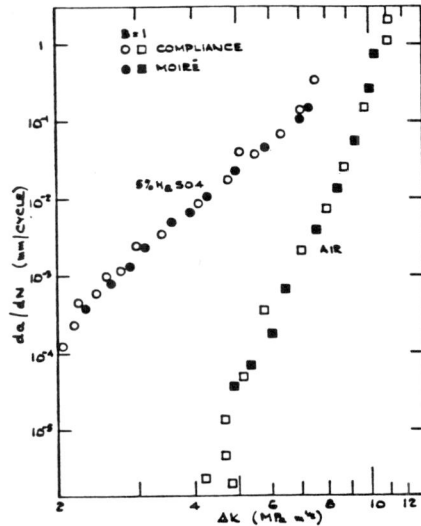

The tests in an air environment so far completed, suggest that the Paris's regime (region II) begins at ΔK equal to 4.5 MPa m$^{1/2}$ and slightly lower for increasing biaxiality B. However, the results obtained in the acidic environment show much lower values, such as 2.5 MPa m$^{1/2}$ or less. This premature fibre damage indicates that the acidic environment enables an early initiation of numerous microcracks with a consequent decrease in fibre strength and cyclic life.

The interesting difference between the two methods of crack measurement is the discrepancy in the magnitude of the crack length. The underestimation of crack length by the compliance technique can be partly explained by the mechanisms of fibre bridging and fibre pull-out, thus holding the crack faces together, reducing the compliance, and therefore the crack length. However, the biaxiality effect is not explained unless stress biaxiality reduces the plastic zone size, on the assumption that the applied biaxial stresses simply modify the uniaxial yield strength of the composite.

PROGRESSIVE DAMAGE DURING FATIGUE

Composite materials are heterogeneous materials characterized by the presence of several types of inherent flaws. These are usually broken fibres, voids in the resin, misaligned fibres, resin-rich zones, or debonded interfaces, the relative presence of each type of defect depending on the manufacturing method used to make the laminate. Therefore, during any deformation process, such as fatigue, the composite material may generally exhibit a variety of failure modes, including debonding, resin cracking, fibre failures resulting from statistically distributed flow strengths, delamination and void growth. In addition, several of these failure modes are generally present at any given time prior to failure. Figure 7 shows the stress strain curve for the composite (c) and for the polyester matrix alone (m) showing that certain defects were generated at relatively low stress levels. This suggests that there are many different ways in which a composite material can become unable to perform its primary function adequately ; thus in each case loss of load carrying capacity or failure can be considered to have occurred. The typical sequence of failure modes leading to tensile fracture in a randomly oriented glass-fibre polyester specimen (see Figs. 8a and 8b) was :

(1) debonding of glass fibres from the matrix

(2) slight loss of structural cracks

(3) prior to failure, a rapid increase in fibre failures leading finally to coalescence of fibre flaws by transverse cracking followed by interfacial shearing and fracture

(4) total fracture momentarily delayed by fibre pull-out.

Thus failure can be shown to be gradual or rapid and may or may not be catastrophic in nature.

Using a microscope, the fatigue specimens could be examined by transmitted light during the course of loading, enabling the progress of damage to be followed. It was possible to observe that individual strands lying perpendicular to the axis of loading were the first site of damage and furthermore, this damage was not particularly associated with the strand ends, but could commence anywhere along the strand spreading in both directions.

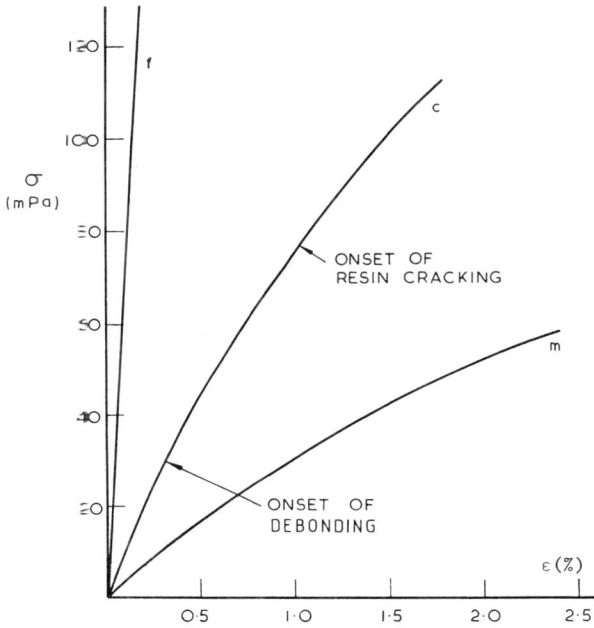

Fig. 7 Average stress-strain curves showing matrix and composite moduli for a strain rate range of $.015 < \dot{\epsilon} < .03$ min^{-1}.

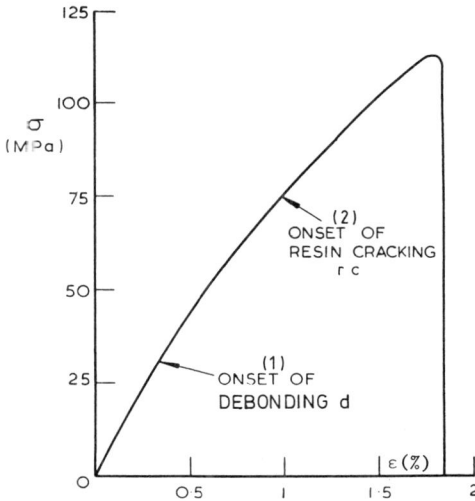

Fig. 8a Strain-stress curve for the composite specimen.

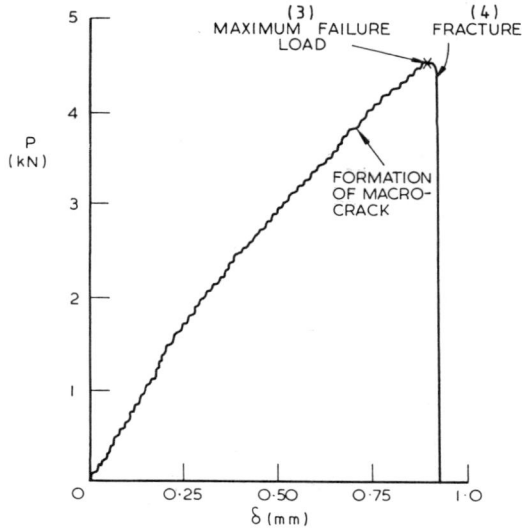

Fig. 8b Load-deflection curve for a composite specimen at a strain rate of 0.015 min^{-1} and room temperature.

Damage sites had a preference for fibre bundles or for those areas where fibres were in close proximity. The first damage took the form of separation between the fibres and the resin matrix. This behaviour is called debonding. As the load was increased, debonding damages was intensified by progressively affecting fibres at smaller angles to the applied load. For a number of tests, debonding occurred from about 25 % to 35 % of the ultimate tensile strength (or K_c) where damage occupied the bulk of the volume of the composite.

At some higher load, less than the ultimate, resin cracks occurred in the resin rich zones of the material. These cracks were also predominantly perpendicular to the line of the applied load and, judging from the appearance of the specimen edges, appeared to originate from some of the debonded areas. Wherever a strand aligned with the loading axis passed through a resin crack, fibre debonding occured on either side of the crack, although apparently without fibre fracture. Again, for the same number of specimens resin cracking occurred from between 65 % and 85 % of the ultimate tensile strength. Just before complete failure, the composite suffered a reduction in load to a varying degree due to localised deformation and damage growth. As the failure strain of the matrix was high in comparison with that of the composite (Fig. 7) the damage zone propagated by sequential failure of the fibres, followed by failure of the matrix along a rough line joining adjacent fibre breaks. Depending upon the degree of fibre-fibre coupling, determined by the proximity of the fibres and the interface shear strength, a break in one fibre was more or less able to influence the point of failure in an adjacent fibre. However, only a certain proportion of fibres was aligned in the direction of the applied load and because the other fibres were discontinuous and oriented randomly, fibre fractures were not always adjacent, and there were many regions of shearing contributing to the work of fracture. Another feature of the progressive crack growth was delamination, where there was not only a large stress concentration tending to open the resin cracks further in their own planes, but also subsidiary stresses tending to open cracks on planes normal to the plane of the crack. Thus, if the composite had planes of weakness parallel with the direction of the applied load, cracks tending to propagate across the material under the influence of this stress would be deflected into non-dangerous paths along the plane of weakness. The delamination process has the effect of arresting cracks and thus momentarily delayed fracture ending with the complete loss of load carrying capacity in the composite after which separation occurred. A trace of a load-deformation graph (Fig. 8b) for a typical low cycle fatigue process discussed here shows that final fracture did not have a pure brittle nature and that some energy was dissipated by shearing and fibre pull-out.

It is realised that the failure of composites is very complex. However, on the basis of these tests the fracture condition can be stated briefly as follows. The ultimate (tensile) strength of the composite depends mainly on voids and other defects as well as on the stress concentration. The effect of the randomly dispersed voids and defects would thus account for the scatter in the toughness results. It is also thought that the composite has reached its ultimate strength when the matrix or the fibres have ultimate strain or the fibre-matrix interface shear strength has the ultimate shear strength i.e., debond strength.

CONCLUSIONS

Low cycle biaxial fatigue of GRP in chemical environments is of considerable interest in structural engineering. The following conclusions were arrived at :

(1) The crack growth rate was investigated in air and acidic environments using two different methods of crack measurement. Despite the variation in crack length magnitude between the moiré and compliance methods, the fatigue crack growth data showed good agreement.

(2) A very high value of the exponent m = 15 in the Paris's crack growth law da/dN = const. ΔK^m suggests a fast crack propagation in air.

(3) However, relative to the results in air, a much slower crack propagation was observed in the 5 % H_2SO_4 environment, where the exponent m decreased to 5.

REFERENCES

1. Wachnicki, C.R. and Radon, J.C., Proceedings ECF2, Darmstadt, VDI, Dusseldorf, pp. 36-64, 1979.

2. Leevers, P.S., Radon, J.C. and Culver, L.E., Polymer, vol. 17, pp. 627-632, 1976.

3. Owen, M.J. and Rose, R.G., J. Phys. D : Appl. Phys., 6, pp. 42-53, 1973.

4. Guerra-Rosa, L., Branco, C.M. and Radon, J.C., Int. J. Fatigue, 6,
    pp. 17-24, 1984.

5. Paris, P.C. and Erdogan, F., ASME, J. Basic Engng., 85, pp. 528-534,
    1963.

6. Radon, J.C., Advances in Fracture Researches ICF 5, Pergamon Press,
    pp. 1109-1126, 1981.

7. Radon, J.C. and Wachnicki, C.R., Proc. Int. Conf. on Fracture Mechanics
    Technology, Melbourne. G.C. Sih, N.E. Ryan, R. Jones, Edts., M. Nijhoff
    Publ., The Hague, pp. 623-634, 1983.

8. Radon, J.C. and Wachnicki, C.R., Biaxial fatigue of fibre reinforced
    polyester resin. Int. Symp. on biaxial/multiaxial fatigue, ASTM,
    San Francisco, California. (In print).

9. Moghadam, S.P., Balthazar, J.C. and Radon, J.C., Int. Conf. on Fracture
    Prevention in Energy and Transport Systems. Rio de Janeiro, 1983.
    (In print).

DEFECTS AND DAMAGE GROWTH DETECTION IN FIBRE COMPOSITE LAMINATES

K. GAMSKI

Civil Engineering Institute, University of Liège, Belgium

SUMMARY

The aim of this contribution is to discuss problems dealing with production defects and its interaction with damage growth and the service life
of composite. The distinction is made between production defects and damage bring about from cyclic loading and environment. A special attention
is paid to the bending process involved both in service life and in testing.

INTRODUCTION

Let me introduce the problem and say that the use of fibre composite materials in cyclic load carrying structural members has raised, from economic
point of view the essential questions of their durability. Although the
fatigue resistance has been studied for years in every aspect as far as
concerning metals, it seems that the extension of what we learned from fatigue research in metallic materials should not be easily exploit to fore-
cast the lifetime of fibre composite materials. In fact the structure of
fibre composite materials is so different that the origine and growth of
damages and delamination development take very specific aspects. Nevertheless the number of cycles to failure N_f is retained as fundamental parameter in composite material fatigue testing. Is n the cycle number at any
time of fatigue test, a nondimensional variable n/N_f may be choosen as a
parameter connected with fatigue damage initiation and growth.

There are many methods which allow to get the elastic or viscoelastic
properties of composite fibre laminate if the properties and proportion of
both fibres and matrix are known. But when we have just properties of
fibres or failure mechanics of fibres and properties of matrix or failure
mechanism of matrix, we have neither the matrix-fibres interaction through
their interface nor the properties of this interface. In general a fatigue
test in view to follow the damage initiation and its growth is lead by
traction, bending, compression or torsion at constant stress amplitude
sinusoidale in shape.

The maximum and minimum stresses fully reversed or zero-tension are choosen refering to the static strength so that damage initiation and its growth
are depending of the difference max-min stresses as well as on the numerical value of each of them. Further the load direction is of importance for
orthotropic fibre composite laminates. Now the damage growth is also very
sensitive to the initial state of material sometimes called "damage free

state". The description of the actual damage free state depends on the investigation method used which may be submicroscopic, microscopic or macroscopic one. Of course the fibre composite laminate comprises always some production defects as air or gas bubbles, interface discontinuities, solid inclusions or matrix shrinkage hairline cracks which may be detected on the surface or on a cross section using adequate lighting and magnifying-glass. The stress concentration or stress intensity factors at these points lead to the premature damage initiation under cyclic load and to the faster and abnormal damage growth. Therefore the fatigue test is extremely useful to control the quality of a production process. But before to start the fatigue test the fibre composite laminate should be scrutated for the number, the nature and the dimensions of production defects. By this way it is possible to define some criterion of the process quality. In this aspect some nondestructive characterisation can be usefully employed as - visual methods - fluorescent penetrants - fluid absorption - ultrasonic both wave speed and signal attenuation - radiographic auscultation - gas permeability.

The visual examination using the surface replica observations gives very interesting results. As for destructive tests the normalised residual strength after a cyclic number of loading can be used.
Fatigue cracks length, void density growth, temperature increase during the cycles testifies also damage growth in fibres composite laminate and may be used as nondestructive tests.

DAMAGE GROWTH IN TESTING

A number of damage parameters can be choosen to examine the damage growth : total matrix crack length, normalized residual strength, number of debonded fibres, for example.
Other damage parameters have been reviewed by J.T. Fong (1) who used some new techniques for detection and monitoring of fatigue damage.
J.T. Fong proposed a conceptual definition of fatigue damage devoted to selection of measurement techniques and parameters for correlating damage and damage growth with fatigue life. Of course he defined a scalar quantity Δ, called damage parameter, so that $\Delta = f(x)$, $f(o) = 0$ and $f(1) = 1$. For convenience Fong has suggested in these criteria the existence of an initial state saying "damage free".

So for $f(o) = a$, $f(1) = 1 + a$, where a is some measurable quantity characterizing the initial damage state of examined composite.
Thus $1 + a$ should be a final damage state or net amount of damage a composite can sustain during service.

$n/N_f = x$, Δ is a state variable which changes with x. At first approximation Fong admitted that the rate of damage growth $d\Delta/dx$ varies linearly on Δ: $d/dx = K\Delta + k$, where K and k are two constants of which only one is available as free parameter to fit the experimental data.
Thus for $K \neq 0$, $\Delta = f(x) = e^{Kx} - 1/e^K - 1$.

If now Δ/n is the amount of damages/cycles and $K \to 0$, the plot of damage parameter $0 \leqslant \Delta \leqslant 1$ versus cycles ratio $0 \leqslant x = n/N_f \leqslant 1$ is straightline (fig. 1). For $K < 0$, Δ/x is convex and for $K > 0$, Δ/x is concave. This modelling method allows to appreciate the regime in which a choosen damage parameter is more sensitive but remains consistent.

153

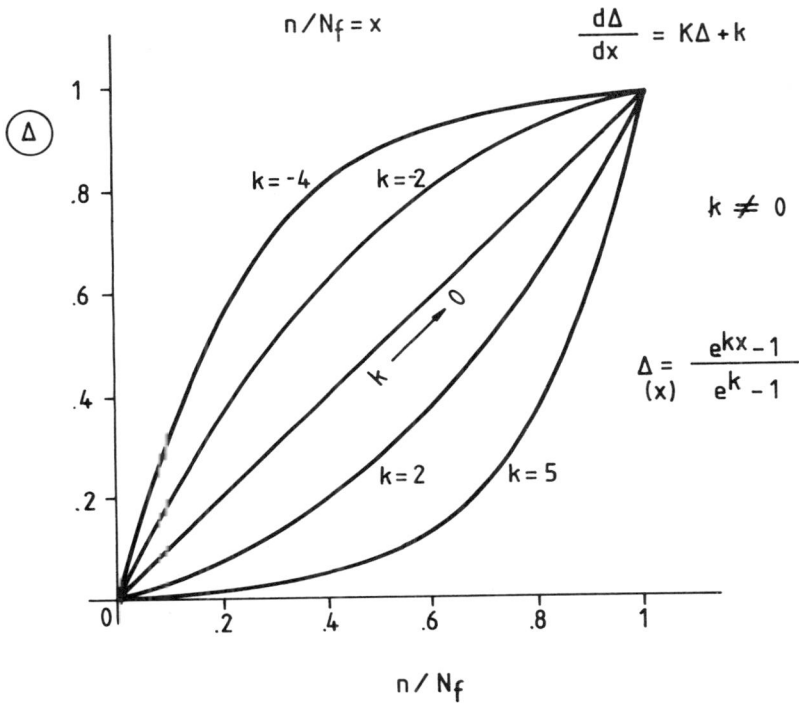

Fig. 1   Δ versus n/N_f for $-4 \leqslant K \leqslant 5$

PRODUCTION DEFECTS IN DAMAGE FREE STATE

From the papers presented in ASTM STP 775, K.L. Reifsneider would have to conclude that fatigue damage consists of matrix damage, that is, matrix cracking parallel to the fibres and delamination. So it is fair to say that matrix cracking and delamination appear to be the types of damages modes closely associated. The matrix cracking is a precursor of delamination.

On the other hand Fong showed that, during the damage incubation period, the residual tensile strength seems to be a sensitive damage parameter which fails a consistency criterion owing the large scatter of results in that regime. After Fong the total matrix crack length is the most promised damage parameter for modelling damage growth.

When the measurements of this parameter are made using X-rays for example,
it seems possible to appreciate the internal discontinuities or production
defects. But when only the surface optic method is applied, what about
eventual production defects ? Maybe that the specimen surface, where the
resin (matrix) thickness is important, is the most vulnerable to the cracks
evolution. It is also easier to scrutate the test specimens as well as
the produced elements on its surface than for their interior. Thus much
is certain that the often used static test is not sensitive enough to detect
a defect except a very large macroscopic one.

In 1959 we observed that for a ductile polymer matrix as PELD, a such
large defect as a hole in the tube wall cross section does not reduce the
tube strength submitted to the axial traction test, but the elongation
at the rupture was dramatically reduced. This
ductile matrix to a brittle one. This stress concentration effect is well
known and carefully studied when steel is concerned. It is important to bear
in mind that the fiber reinforced composite is made of very different
materials regarding their mechanical properties.
Its continuity is depending on adhesion between matrix and reinforcement
and so a weakness in adhesion should be carefully avoided. The debonding
can easily be seen on the bending of currugated sheets. After some cylces
the delamination is clearly seen on the compressed waves.
The visual, eventually using magnifying glass, examination allows to class
some production surface defects in "damage free state" : lack of impregnation - crazing - shrinkage cracks - fibres emergency - delamination or
lack of adhesion (photographs 1 to 5).
Other defects can be discovered on the transversal cross section, injection
voids, inclusions, delamination.

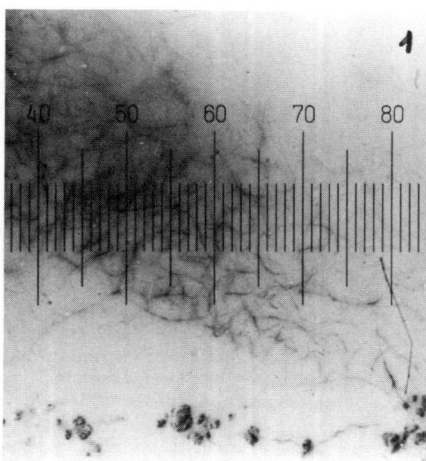

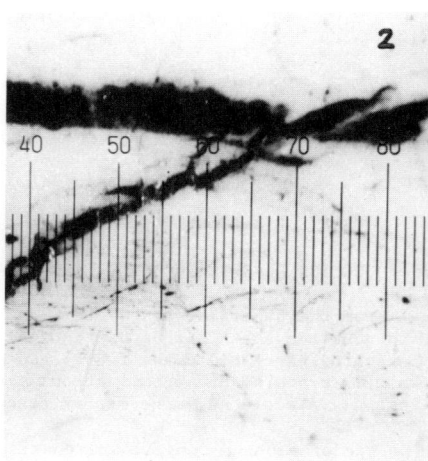

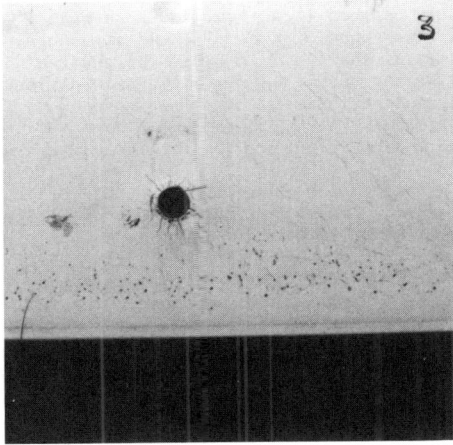

DAMAGE DEVELOPMENT IN STRUCTURAL ELEMENT AND IN FATIGUE TEST

We are all aware that the ultimate goal of fatigue test in its various aspects is to be able to predict N_f with confidence. But fatigue damage modelling of some complicated in shape structural element is costly and time consuming. Therefore fatigue tests are lead on the simple samples rather then on a whole structure piece, although in some special cases, the fatigue tests are also undertaken on whole structural elements.

Nevertheless it is extremely difficult to reproduce in laboratory a complicated set of physical and mechanical charges which happen in situ. So the
question is raised, when starting from fatigue tests lead on some elementary
specimens submitted to a unique set of charges (traction, compression,
torsion), how to predict the fatigue resistance of a complicated structural
element submitted to a complicated set of charges. At first the charges
used in fatigue tests are generally higher than those in situ, secondly the
cycles are not of the same intensity and periodicity.
For want of something better it is extremely useful to scrutanize, in situ,
the behaviour of structural elements and to compare it to the fatigue damage
development observed in fatigue test
The damage observations made in situ lead to the conclusion that the cracks
apparition is followed by hair line cracks development and fibres emergence
(local). The shrinkage stresses in combination with bending leads to coin
cracks or cracks starting from perforations.
Under bending, stress cracks develop between two defects of impregnation.
The fatigue tests were undertaken on test specimens in pure polyester resin,
polyester with mineral charges, polyester reinforced with mats and woven
rovings in various combination and on thermoplastes and fibres reinforced
thermoplastes.
A special developed testing device allows to test simultaneously six samples
in bending when clamped at one end and pinned-up to a vertically moving bar
at the second end. So the sample works as a cantilever beam and its shape
is adapted to have a max. stress in a zone near the pinned-up end. All samples
are submitted to a constant deflection for 30 Hz, (fig. 2.a).

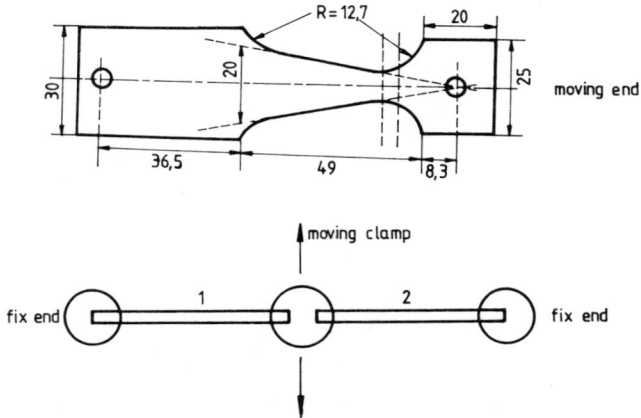

During the fatigue life the sample surface may be scrutanized for cracks
development. This method has some advantages but also some inconveniences.
This device is easy to construct and cheap, it allows to work quickly
(six samples together). A calibration using classic dynametrid device
allows to precise the force in action. The deflection amplitude is variable
through tie-road and crank. Each sample is connected with a cycles counter.
The fatigue cycle is a fully reversed one.
In view to illustrate the reached results, as concerning fibre composite
laminated these obtained on the polyester, reinforced with woven roving.
are showed on the annexed (figures 3, 4).

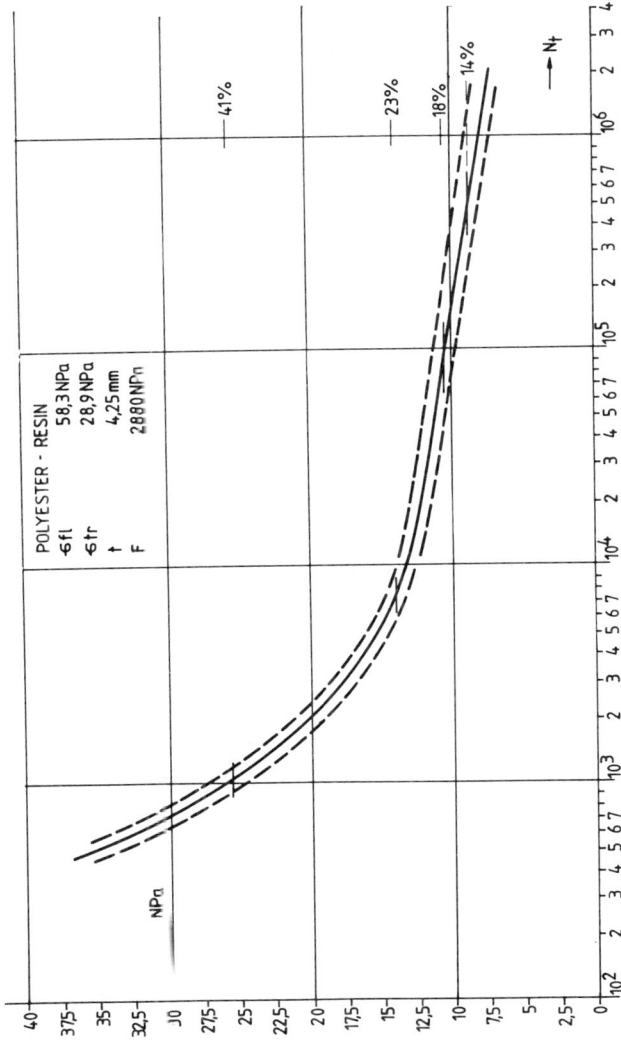

Fig. 3

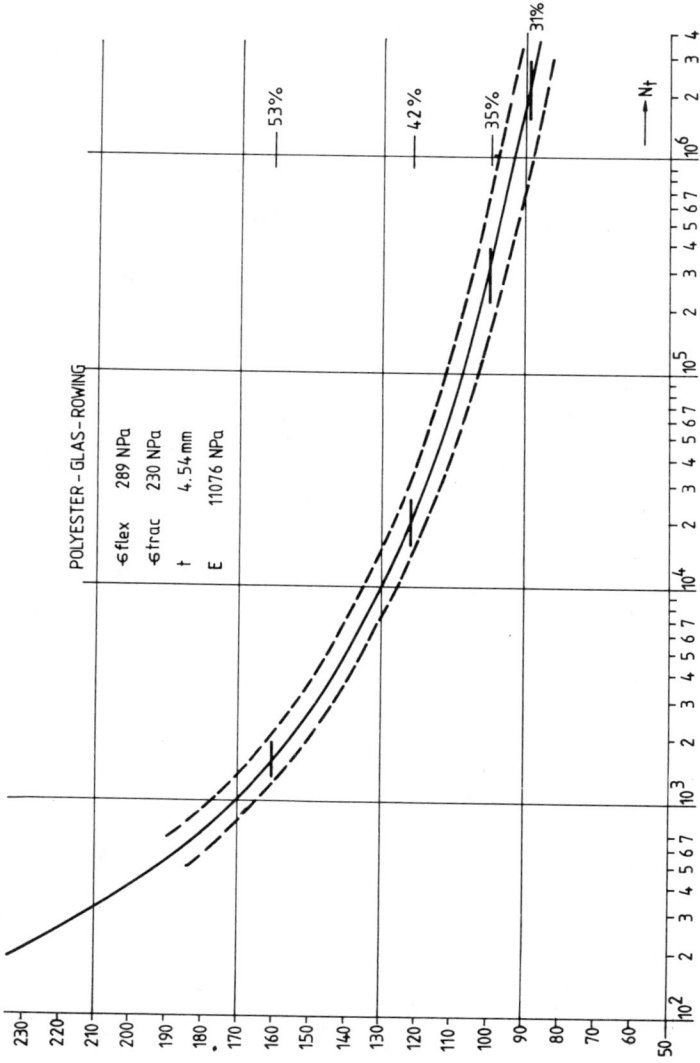

POLYESTER – GLAS – ROWING

σflex   289 NPa
σtrac   230 NPa
t   4.54 mm
E   11076 NPa

53%

42%

35%

31%

$\longrightarrow N_t$

Fig. 4

CONCLUSIONS

Laminated composites are built of two or more materials very different in nature as their mechanical properties is concerned. The elastic modulus of matrix polymer is 15 - 70 times lower than this of fibres as it can be seen on the table below.

material	tensile modulus (GPa)	tensile strength (GPa)	density
E-glass	73	3,5	2,54
S-glass	86	4,5	2,48
Thornel	200-550	1,5-2,7	1,4-1,3
Graphite HE	390	2,1	1,9
Graphite HR	240	2,5	1,9
Silica	73	5-8	2,2
Steel	210	1,5-2,1	7,8
Tungsten	415	4,2	19,8
Kevlar 49	130	2,8	1,5
Epoxy/Polyester	5	0,01	1,2

Bonding of the fibres and protect them in various environmental conditions are the essential parts the matrix has to assume in composite. The relaxation time is in polymer matrix many times higher than in fibres. Thus the stress distribution in composite cross-section between fibres and matrix is very different in static tensile test, in static low strain rate tensile test as well as in creep the stress is, of course after a transition period, transferred to the fibres (4). W.S. Carswell et al. (5) showed a relationship existing between the tensile creep rupture properties at room temperature on polyester and epoxy laminates reinforced with two types of glass cloth, and at elevated temperatures on the polyester laminates, reinforced with satin weave cloth. Damage as seen by optical microscopy, appeared to be confined to isolated regions of delamination between fibres and matrix until first failure of the fibres produced localized matrix cracking. It must be underlined over and over again that in this case the total load in a cross-section is supported by fibres following the relaxation of the matrix. The individual fibres do not support the same stress ; for the technological reasons some fibres are bend or foild so that their initial length is not the same and so neither the initial stress. That is the reason why the fibres failure is progressive. In fatigue or under fast cyclic loading the stress relaxation in matrix can not operate.

The elastic moduli ratio being E_F/E_M = 15 - 70, there is a large stress difference between fibres and matrix, so that the matrix-fibres interface has to support a high shear stress. That is the reason why in these conditions some localized matrix cracking is first observed.

160

The cross-section being weakened the adjoining fibers are overloaded and
their rupture follows.   Anyway a production defect as delamination,
porosity, air bubbles, accelerate the failure of both under constant and
in cyclic loading.   Are these production defects small in size, they
would be hardly ever seen in a simple tensile test but they could become
a starting point for cracks growth.
The ultimate goal of fatigue tests is to be able to predict for a given
structural component the cycles number at failure N_f.
In view to exploit the sample - N_f to predict structural component - N_f
it should be recommended to get a correlation between damage development
of a sample under cyclic load and damage development of a structural component,
observed in situ, and submitted to the actual loading.

REFERENCES

1. Damage in composite materials. ASTM STP 775, 1980.

2. Gamski, K., Effets des défauts de la paroi des tubes en PE,
    Matières Plastiques, Brussels, 1959.

3. Review of developments in plane strain fracture toughness testing.
    ASTM STP 463, 1970.

4. Gamski, K., Considérations sur le comportement des stratifiés soumis
    à des sollicitations mécaniques de courte et de longue durée.
    Alumni-Amici, Université de Liège, 1964.

5. Carswell, W.S. et al., Creep rupture and tensile tests on glass-
    reinforced laminates. NEL report 439, Glasgow, 1969.

ANALYSIS OF FATIGUE DAMAGE IN CFR EPOXY COMPOSITES BY MEANS OF ACOUSTIC
EMISSION : PREPARING A CALIBRATION EXERCISE

M. WEVERS, I. VERPOEST, E. AERNOUDT and P. DE MEESTER

Catholic University of Leuven, Belgium

ABSTRACT

The degradation of the mechanical properties of carbon fibre reinforced
epoxy composites caused by fatigue is due to the development of specific
damage modes. The acoustic emission technique was found to be very useful
in monitoring these different damage modes on a time resolvent and strictly
passive way. Other techniques however, such as the replication technique
and X-ray radiography were needed in order to get a calibration of the acoustic emission signals. The calibration tests are outlined in this paper.

INTRODUCTION

The carbon fiber reinforced epoxy composite is finding an increasing use
in structural applications because of its high specific strength and stiffness. However the reliability of these structures depends to a large extent
on the ability to withstand cyclic loading which therefore must be carefully
investigated.
It is known that the degradation of the mechanical properties of a cfrp specimen during fatigue tests, such as strength reduction and loss of stiffness, is due to the development of specific damage modes in the material :
matrix cracking, debonding between fibre and matrix, delamination between
plies and fibre fracture (Jamison, Schulte, Reifsnider and Stinchcomb 1982,
Poursartip, Ashby and Beaumont 1982, Reifsnider and Talug 1980).
The detection and identification of these damage modes during fatigue testing
is necessary in order to estimate the residual strength or stiffness of the
specimen since different types of damage can effect the strength or stiffness
differently.
In order to identify the damage modes in a composite material non destructive
testing techniques such as ultrasonic attenuation measurements (C-scan) and
radiography can be applied (Sendeckyj, Maddox and Tracy 1978). The acoustic
emission (AE) is based on the detection of elastic surface stress waves
caused by the dissipation of elastic energy. Plastic deformation as well as
crack initiation and growth can be sources of these elastic waves, and hence
AE is also used in monitoring fatigue damage. The advantage of this NDT-
technique, in contrast with those mentioned earlier, is that it can monitor
the damage on a time resolvent and strictly passive way. The AE-signals are
retrieved while fatigue loads are being applied without influencing the
damage development. The AE-technique was used in this study to obtain information about the basic fracture mechanisms and to detect and identify the
different damage types.

Each failure mode can be expressed to release a different amount of energy and hence to generate different amounts of AE. Several researchers (Mehan and Mullin 1971, Becht, Swalbe and Esenblaetter 1976, Wolters 1982 and Williams 1980) have already studied the failure mechanisms in fibre reinforced composites with the acoustic emission technique by observing the amplitude of the AE-signals.

In this study on carbon fibre reinforced epoxy composites the detected and accepted AE-signals were analysed on the basis of the energy content of the signal. The gathered information was compared with the observation of the replicas from the side surface of the specimen at certain time intervals and an X-ray radiograph of the specimen at the end of the test. By combining all these techniques it was possible to characterize the AE-signals from different damage types : matrix cracking, delamination, debonding between fibre and matrix and fibre fracture.

EXPERIMENTAL METHODS

Materials

The cfrp test specimens used in this work were cut from plates made by heated press technique by N.V. Snauwaert. These plates were made from 8 layers cut from commercially available preimpregnated sheets : Fibredux 920C-TS-5-42 by Ciba Geigy. Two different laminate types were chosen for this study : unidirectional $(0°)_8$ and cross plied $(+45°,-45°,+45°,-45°)_{2s}$ laminates.

Mechanical testing

The fatigue tests were carried out on a servo hydraulic fatigue testing machine Schenck Hydropuls 25 kN.
The $(0°)_8$ laminate was stressed up to σ_{max} = 726 MPa (70 % of the UTS) with $R(\frac{\sigma_{max}}{\sigma_{min}})^{-1}$ = 0,03 and the $(+45°,-45°,+45°,-45°)_{2s}$ laminate up to σ_{max} = 58,28 Mpa (40 % of the UTS) with R = 0,03 and K = 1,46. The test frequency is 3 Hz , and the tests were carried out in an air-conditioned room at 20 ± 2°C and 45 ± 5 % relative humidity.

Acoustic Emission equipment

An AET 5000 computerized emission monitoring system was used in this study. When a specimen is loaded, acoustic emissions are detected by two transducers in order to enable linear location. These transducers have a flat frequency characteristic in the 200-550 KHz range. After preamplification (x 60 dB) and filtering the AE-signals are processed by the signal processor unit which can add a variable amplification (up to 40 dB) to the signal. The microcomputer provides data acquisition, handling and function analysis. The test data are analysed on-line and graphic representation is available. Only the "good" events are processed and displayed on graphs : i.e. acoustic emission signals :

- reaching above a preset threshold
- within a linear location area
- accepted by the data discrimination schemes : i.e. the characteristics of the acoustic emission signal, the amplitude, energy, event duration, number of ringdown counts, slope and rise time have to be greater than zero and smaller than a preset value.

The AE-signals were mainly followed up by their location between the two transducers and their energy.

Non-destructive testing

Two supplementary techniques were used in order to calibrate or identify the AE-signals : the replication technique and X-ray radiography. Edge replication of composite specimens has been employed with success in the study of ply cracking, edge delamination and even fibre fracture for the $(0°)_8$ laminate. In order to use the technique for composite materials, the specimen edges should be polished.

A piece of cellulose acetate tape, 100 μn in thickness for the cross plied laminate and 34 μm in thickness for the unidirectional laminate, is doped in acetone from one side and pressed quickly and lightly against the specimen edge. The acetate tape is allowed to harden for approximately one minute and upon removal it retains a permanent, high fidelity replica from the edge topography. The investigation of the replicas was done first with a micro-fiche reader (magn. 24x) and then successively with the optical light microscope and the scanning electron microscope after coating them with a thin conductive film of gold.

At the end of each fatigue test an X-ray radiograph was taken in order to visualise the internal damage state of the specimen. Because the $(0°)_8$ laminate was not broken we used tetrabomoethane (TBE) as an enhancing agent. This penetrant exhibits excellent X-ray opacity and diffusivity in carbon-epoxy material but care must be taken for this penetrant is quite toxic. The radiographs were etaken using a Balteau X-ray unit (5-50 kV, focal spot 100 μm). The nominal film-to-source-distance was 55 cm. The used operating voltage was 25 kV and the source current was approximately 5 mA. The film for the contact radiography was an Agfa-Gevaert D4-type. For these operating conditions the exposure times were 35 seconds for the $(+45°,-45°,+45°, -45°)_{2s}$ laminate and 1 minute for the $(0°)_8$ laminate.

EXPERIMENTAL RESULTS AND DISCUSSION

Cross plied $(+45°,-45°,+45°,-45°)_{2s}$ laminte

The optical study of the replicas and the X-ray radiography of the cross plied laminate showed that matrix cracking and delamination between the plies were the main damage modes. Matrix cracking did appear first and gave onset to delamination. Evidence of these two damage modes is presented on the SEM-photographs of a replica taken after 150.000 fatigue cycles (Fig. 1.a).
The energy distribution graphs after a certain number of fatigue cycles, were compared throughout the fatigue test with the replicas taken after the same number of cycles. An example of these energy distribution graphs is given in Fig. 1.b taken after 163.000 fatigue cycles.

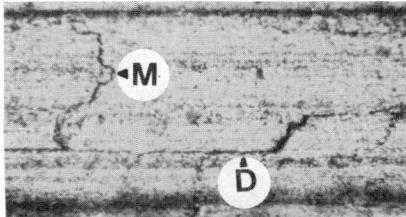

Fig. 1.a : $(±45°)_{4s}$ matrix cracking (M) and delamination (D) - (150.000 cycles)

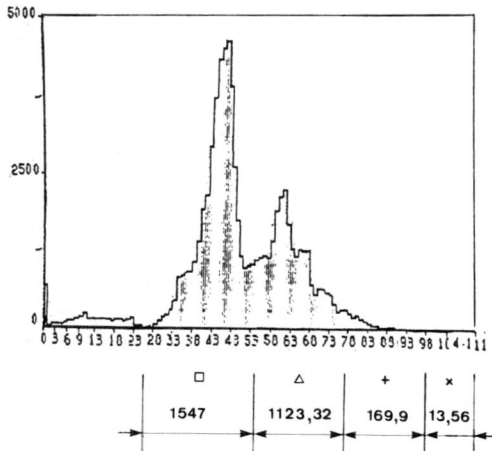

Fig. 1.b : Energy distribution of the AE-signals from the cross plied laminate
(163.500 cycles).

Accounting for the dominant damage modes of this cross plied laminate and
the succession of these modes, we could find that the AE-signals of matrix
cracking are gathered in the [26,54] energy interval and that delamination
emits signals of the [55,76] energy interval. Once these two damage modes
were correlated with intervals of the energy distribution graphs, we could
plot growth curves of the damage modes in function of the number of cycles :
curve 1a, 2a, 1b and 2b of Fig. 7. These curves also indicate that for the
cross plied laminate matrix cracking gives onset to delamination and that
at a certain number of cycles matrix cracking is the most important damage
mode (much more important than delamination).

On the log of the energy distribution graph, AE-events with an energy
content higher than 76 were also visible near the end of the test but we
could not correlate these AE-signals with other damage modes at that moment.

Unidirectional (0°)$_8$ laminate

The first dominant damage mode visible on the replicas of the edge of the
unidirectional laminate was delamination between the 0° plies (Fig. 2).
Later on, other damage modes like matrix cracking, debonding between fibre
and matrix and fibre fracture were also developed during the fatigue test
see Fig. 3. The fibre fractures however only became important after approximately 500.000 cycles and they could best be resolved by the sem-study
of the remplicas (Fig. 4).

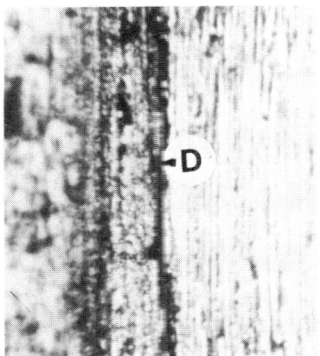

Fig. 2   (0°)₈ Delamination (D) of
          the 0° plies (200.500 cycles)

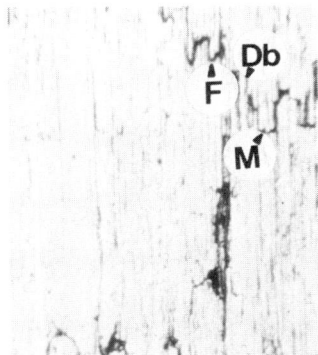

Fig. 3   (0°)₈ Matrix cracking (M),
          debonding between fibre and
          matrix (Db), and fibre fracture
          (F) - (453.600 cycles)

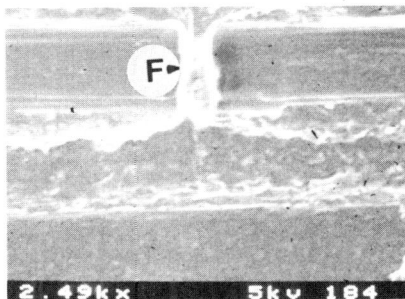

Fig. 4   (0°)₈ fibre fractures (F) - (961.300 cycles)

For this unidirectional laminate we also combined the optical study of the
replicas at certain numbers of fatigue cycles with the energy distribution
graphs of the gatered AE-signals after the same numbers of cycles. This
correlation showed us that for this laminate the AE-signals of delamination
are gathered in the [54,79] energy interval and that matrix cracking emits
AE-events of the [26,53] energy interval. This we previously already learned
from the cross plied laminate. Furthermore the AE-signals of debonding between fibre and matrix and fibre fracture could be correlated with a visual
observation too as the fibres, parallel with the side surface, can be seen
on the replicas. So we found that the 80,98 energy interval gathers the
fractures. The energy distribution graphs after 14.250 and 961.300 fatigue
cycles are presented in Figs. 5 and 6.

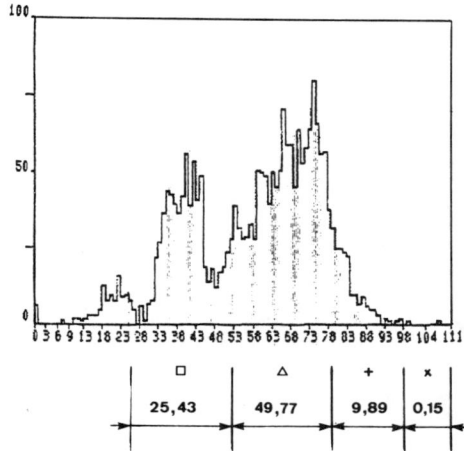

Fig. 5 Energy distribution of the AE-signals from the unidirectional laminate (14.250 cycles)

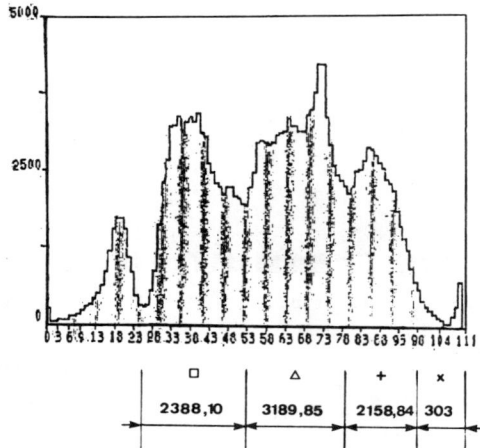

Fig. 6 Energy distribution of the AE-signals from the unidirectional laminate (961.300 cycles)

In analogy with this test the energy interval [77,98] of the cross plied laminate could be correlated with debonding and the [99,111] interval with fibre fracture although these damage modes were less important.

For the unidirectional laminate and the two invisible damage modes of the cross plied laminate growth curves were also calculated (see Fig. 7, curves 3a, 3b and 4b and Fig. 8, curves 1, 2, 3 and 4).

These curves show clearly that :

1. debonding and fibre fracture are the minor damage modes in the cross plied laminate and become important at the end of the fatigue test only

2. delamination is the first developing damage mode in the unidirectional laminate and fibre fracture is important only after 500.000 fatigue cycles

3. the development of delamination (although first), matrix cracking and debonding between fibre and matrix in the unidirectional laminate is steady and almost equally important as also can be observed from the energy distribution graphs, whereas in the cross plied laminate matrix cracking is far more important than delamination.

The energy distribution graphs of the AE-signals are very useful in order to identify the different damage modes in the cfrp composites. In literature the fibre fractures are associated with AE-signals of higher amplitudes than the other damage modes (Becht et al. 1976, Wolters 1982). This is in agreement with the findings of this study.

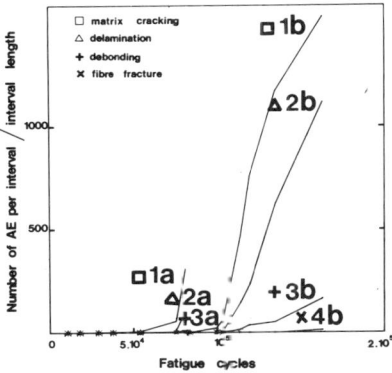

Fig. 7 Growth curves of the damage modes in the cross plied laminate during fatigue

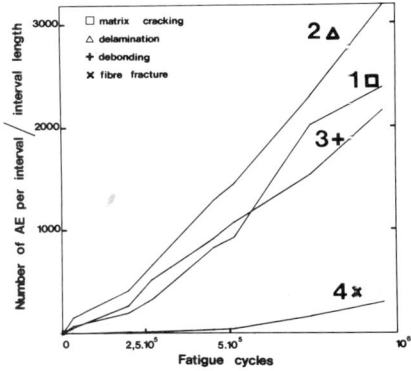

Fig. 8 Growth curves of the damage modes in the unidirectional laminate during fatigue

CONCLUSIONS

The acoustic emission technique was found to be very useful in monitoring the basic fracture mechanism in the carbon fibre reinforced epoxy material. The calibration tests clearly identified the different emission mechanisms from the different damage modes. More fatigue tests should be carried out to make statistical analysis possible. In future work the damage growth curves will be correlated with the stiffness reduction of the laminate in order to set up a damage accumulation theory for carbon fibre reinforced epoxy composites.

REFERENCES

Becht, J., Schwalbe, H.J. and Eisenblaetter, J., Acoustic emission as an aid for investigating the deformation and fracture of composite materials. Comp., pp. 245-248, Oct. 1976.

Jamison, R.D., Schulte, K., Reifsnider, K.L. and Stinchcomb, W.W., Characterisation and analysis of damage mechanisms in fatigue of graphite/epoxy laminates. Presented at the ASTM symposium on the effects of defects in composte materials, San Francisco, California, 13-14 December 1982.

Mehan, R.L. and Mullin, J.V., Analysis of composite failure mechanisms using acoustic emission. J. Comp. Mat. 5, pp. 266-269, 1971.

Poursartip, A., Ashby, M.F. and Beaumont, P.W.R., Damage accumulation in composites during fatigue.
In Proc. of the 3rd RISØ International Symposium on Metallurgy and Materials Science, Fatigue and Creep of Composite Materials. Edited by H. Lilholt and R. Talreja, (RISØ), pp. 279-284, 1982.

Reifsnider, K.L. and Talug, A., Analysis of fatigue damage in composite laminates. J. Fat., pp. 3-11, Jan. 1980.

Sendeckyj, G.P., Maddox, G.E. and Tracy, N.A.
In Proc. of the 1978 Int. Conf. on Comp. Mat., ICCM/2, New York, p. 1037, 1978.

Williams, R.V., The application of acoustic emission to fibre-reinforced materials and to concrete.
In Acoustic Emission, edited by Adam Hilgar Ltd, Bristol, p. 107, 1980.

Wolters, J., Influence of shape and bonding of glass particles in acoustic emission behaviour of polycarbonate (to be published), 1982.

EXAMINATION OF IMPACT RESISTANCE OF FRP - SUGGESTION FOR A STANDARD TEST METHOD

K. STELLBRINK

Deutsche Forschungs- und Versuchsanstalt für Luft- und Raumfahrt, Institüt für Bauweisen- und Konstruktionsforschung, Stuttgart, Germany.

Fibre reinforced plastics, especially carbon fibre reinforced epoxies are susceptible to impact loading, e.g. by hailstones, runway debris, clods of clay or dropped tools or handling shoves.
Since their application increases e.g. in aircraft structures, it is necessary to know, among others, their impact behaviour. This must be acquired by analytical and experimental methods. The elastic and fracture response of composite structures is manifold and obviously can not be discovered by a single impact test. The experimental investigation of impact behaviour seems to be a very expensive affair. This paper proposes how to reduce the experimental cost to minimum.

The most important parameters, governing the impact response of thin composite shells, are :

- impactor velocity, impactor mass, shape of the contact tip
- shape of the shell, support condition
- laminate stiffness, laminate thickness
- fibre/resin type, stacking sequence.

The influence of impactor mass or velocity is low in a certain range, as long as the incident energy, determined by mass and velocity, is kept constant. By comparison, the geometrical relations have a strong influence. As an example, Fig. 1 illustrates the influence of the dimension. It shows the computed load and displacement versus time of two identically loaded, similar CFRP plates. Only the size of the plates is different. The larger one (1 m x 1 m) reveals lower contact forces and larger deflection and contact duration than the smaller one (74 mm x 74 mm). The stress within the plate at indent location are very different, forcing different (fracture) behaviour of the laminate.

Most of these parameters which influence the impact response are unpredictable or scarcely affectable by the designer. Best of all, the material selection can be done with respect to impact response. As a first consideration it is sufficient to know the impact behaviour of a candidate material compared to another similar one.

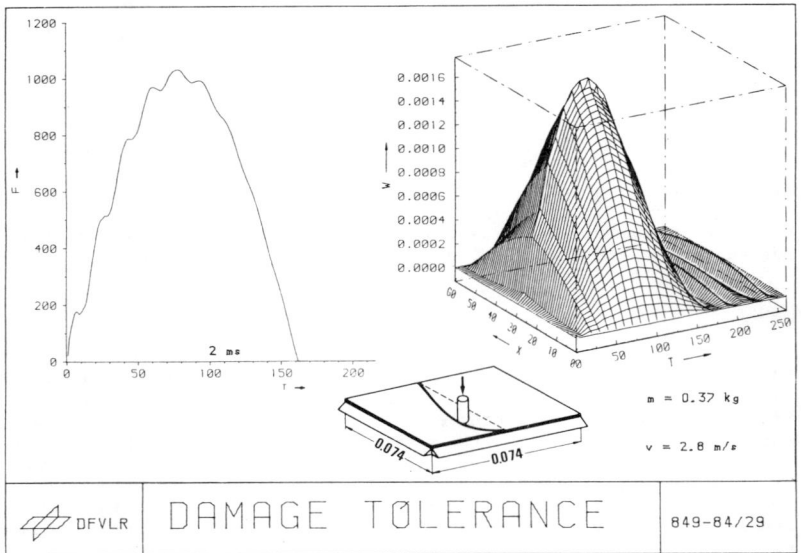

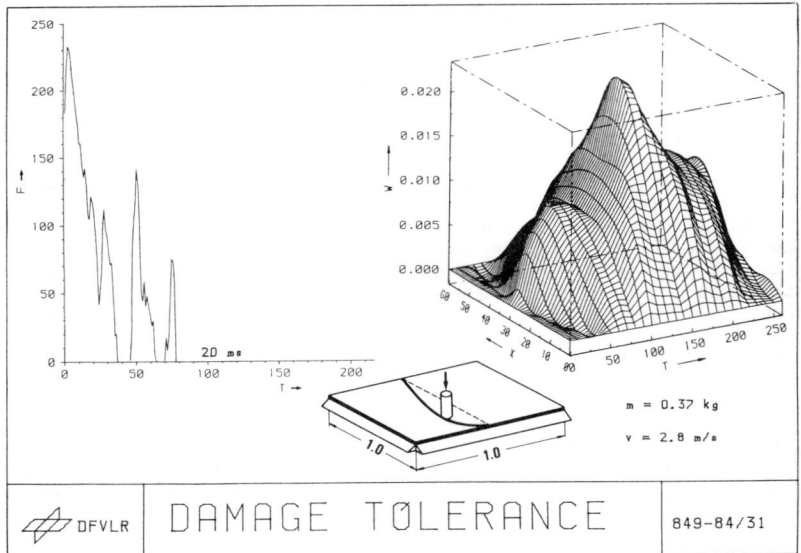

Fig. 1 Computed dynamic response of impact loaded laminate T300/914C
(0/0/±45/90/±45/0/0) symm. ; plotted are load and centre-line
deflection versus time with two different plate sizes.

The macroscopic damage behaviour of an impact loaded composite structure depends mainly on the parameters mentioned above. But the internal microscopic fracture in each known case starts with cracks in matrix and interface, followed by interlaminar delaminations and finally fibre breaks, although in a variable ratio (Fig. 2).

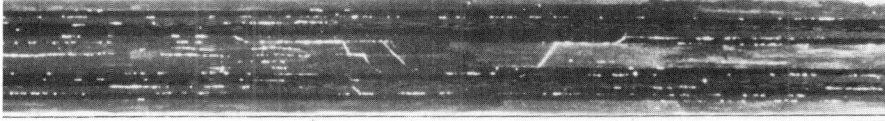

Impact energy E_O = 1.4 J ; clamp diameter D = 160 mm ; hemispherical impactor tip diameter d = 5 mm ; absorbed energy ΔE = 0.56 J.

E_O = 4.8 J ; D = 160 mm ; d = 10 mm ; ΔE = 2.3 J.
E_O = 4.8 J ; D = 160 mm ; d = 10 mm ; ΔE = 2.3 J

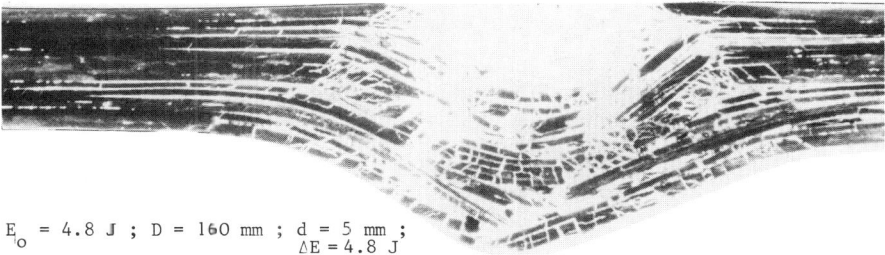

E_O = 4.8 J ; D = 160 mm ; d = 5 mm ;
$\Delta E = 4.8$ J

Fig. 2 Typical damages in CFRP laminates, (0/0/±45/90/±45/0/0) symm., covered with E-glass (style 120).

For ranking the impact resistance of composites we should use a test procedure which allows the observation of all three modes in a fairly balanced proportion. The hypothesis is made, that the ranking of various materials concerning the occurence of different damage modes, is insensitive to impact or test conditions. In other words : if a material reveals fibre breaks earlier (e.g. at lower energies) than a competing material with the identical set of impact parameters, this will be the case with altered test conditions, too.

Accepting this reasoning we could conclude, that a single properly selected impact test would be sufficient. Unfortunately, this is not reasonable, which is illustrated in Fig. 3. Two composites plates, similar with the exception of the matrix, are impact loaded identically, each with two different indent energies, the remaining parameters kept constant. With the lower energy level, a clear ranking of the two resins is shown regarding the projected delamination area, monitored by ultransonic scan. With the higher energy level the clear difference disappeared, which is also true for the corresponding absorbed damage energies.

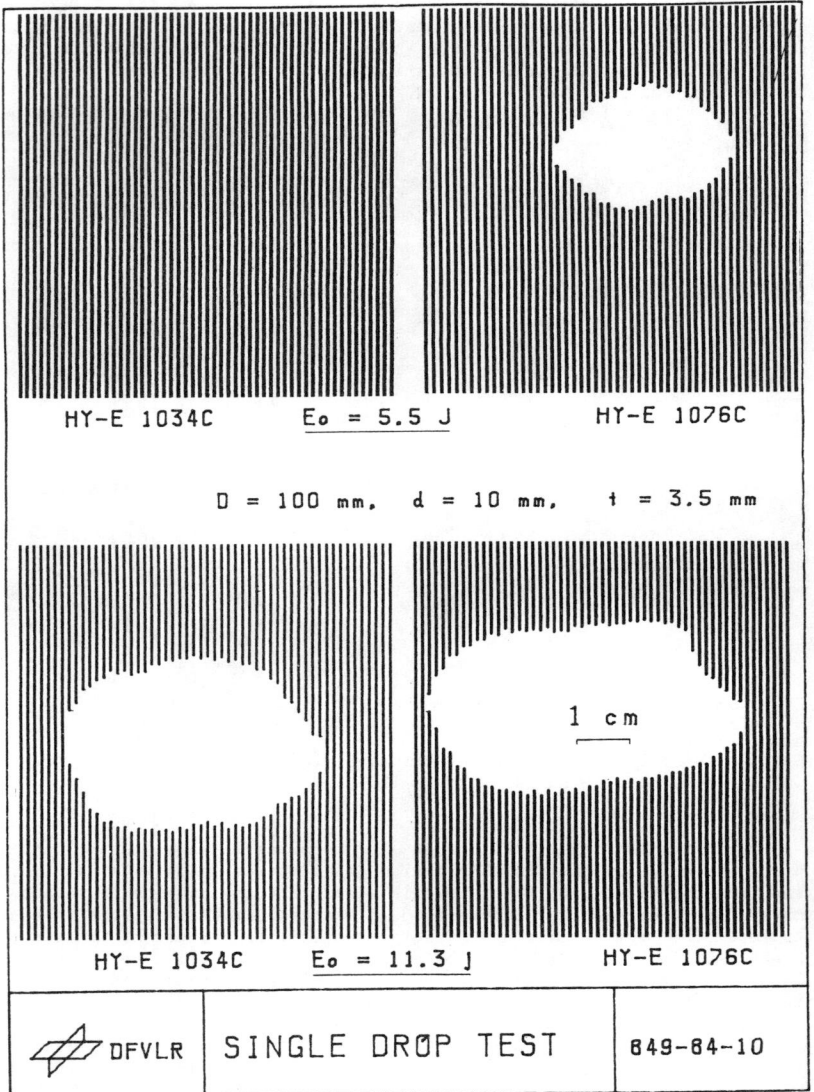

Fig. 3 Interlaminar delamination (us-scan) of two laminates (0/45/0-45/90/-45/0/45/0)3x at two different impact energy levels.

We need test series with smaller and higher impact "power" to produce small damages including fibre breaks to enable comparison of material behaviour at different damage modes. The simplest way to do this is testing at different levels of incident energy. In Fig. 4 some characteristic values, evaluated from an accelerometer on top of the impactor, are displayed. The time of deflection T, the maximum indent force F and the absorbed energy E_{dam}, all indicate the change of fracture modes at incident energy E_o of about 1.5 J and 4 J. The maximum deflection d does not exhibit any characteristic in this clearness. This figure demonstrates that we only need the absorbed energy E_{dam} versus the incident energy E_o in order to find changes of the characteristic failure mode.

But this can be done with a small amount of material. The left four sub-diagrams in Fig. 5 show, in solid lines, similar curves to those of Fig. 4. These curves represent the maximum deflection, duration of deflection, maximum force and absorbed energy of a test series where different specimens were impact loaded with different levels of incident energy. Accompanied are especially marked data points, showing maximum deflection, duration of deflection, maximum force and absorbed energy of a test set where one specimen was impact loaded with different levels of incident energy. The sequence of these energy levels was not arbitrary but started with the lowest energy and continued with the next higher one until penetration occurred. The energy steps were constant. In other words : a weight dropped onto a thin composite plate from a drop height of 50 mm, then from a drop height of 100 mm, then from a drop height of 150 mm, etc. Until final rupture. Except with the maximum force, the test data of the Repetitive Dropweight Test with Increasing Energy (RDTIE) display a similar shape than those of the single drop test, but compressed along the incident energy axes. The RDTIE looks somewhat like a condensed single drop test, supplying similar cognitions. With this test it is beneficial to use a different graphical presentation. The right subdiagram in Fig. 5 shows the accumulated absorbed energy over the accumulated incident energy, forming a quasi curve of discrete data points. This kind of data representation has the advantage, that the scatter of individual test data are leveled out by accumulation, but systematic differences between two specimens and enhanced are easier to detect than with single drop test series.
Fig. 6 shows an application of the RDTIE. It compares the impact behaviour of E-glass and T3CO carbon fibre in the same matrix and arrangement. It demonstrates clearly the superiority of glass fibres. GFRP absorbs more energy and sustains higher impact loads than CFRP does. In Fig. 7 the impact behaviour of some carbon fibre/epoxy systems is compared. Each system was cured at 175°C and was declared impact resistent by the manufacturer. The diagram confirms that both, high strain fibres and toughened matrix system are needed to produce good damage behaviour. Unfortunately, these toughened matrix systems have worse properties under hot/wet service conditions.

The repetitive dropweight test with increasing energy is a simple test. It puts us in a position to rank composites with respect to absorbed and sustained incident energy up to penetration. Still we can rank the appearance of different damage modes, matrix/interface cracks, delamination, fibre breaks. In Fig. 8 this distinction can be made by virtue of the slope of this pseudo curve. There are plotted the load and deflection versus time for three selected drop tests together with the corresponding energy data.

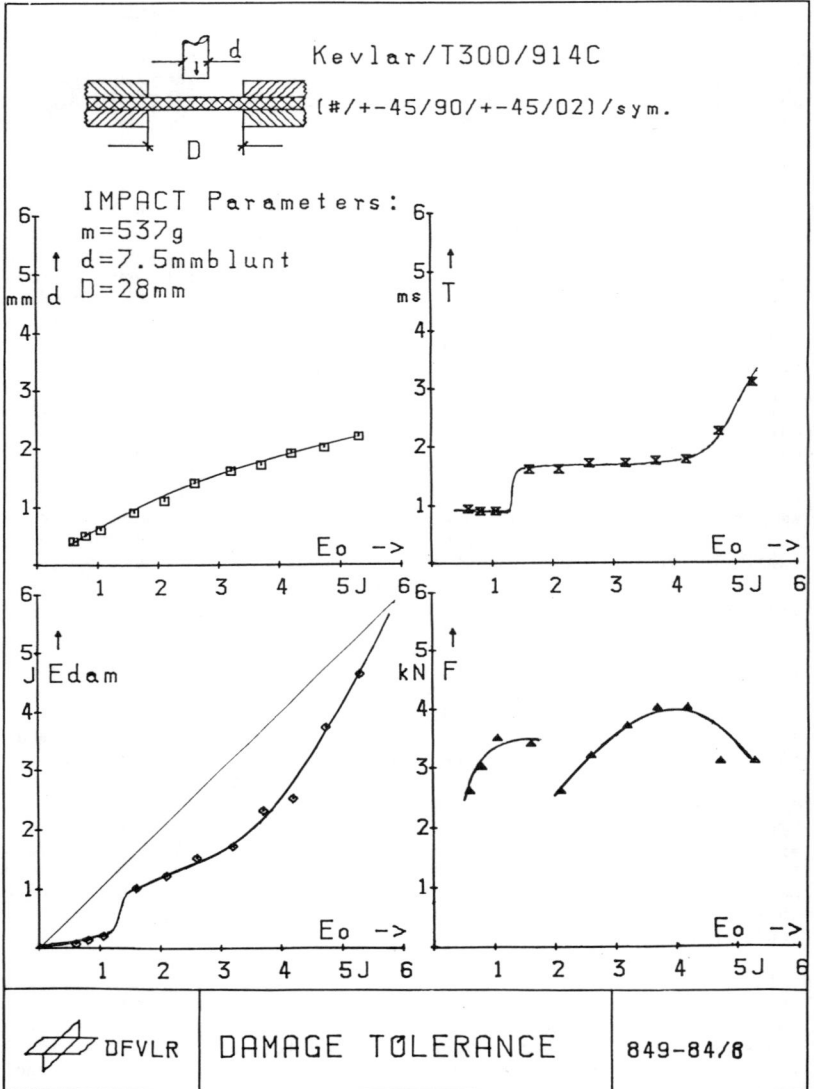

Fig. 4   Impact test data of hybrid laminate (Kevlar styme 181 and T300 warp
sheets), demonstrating the reflection of different damage modes in
the test data.

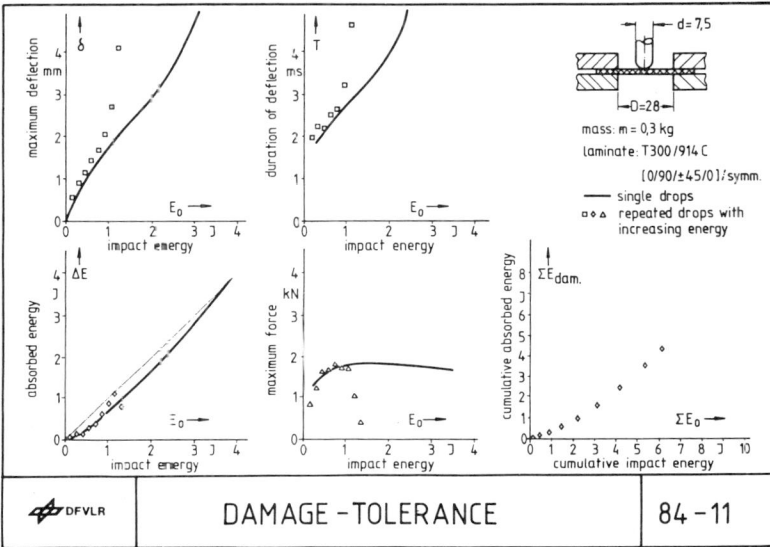

Fig. 5 Comparison of test data of a single drop weight test set and a repetitive drop weight test with increasing energy (RDTIE).

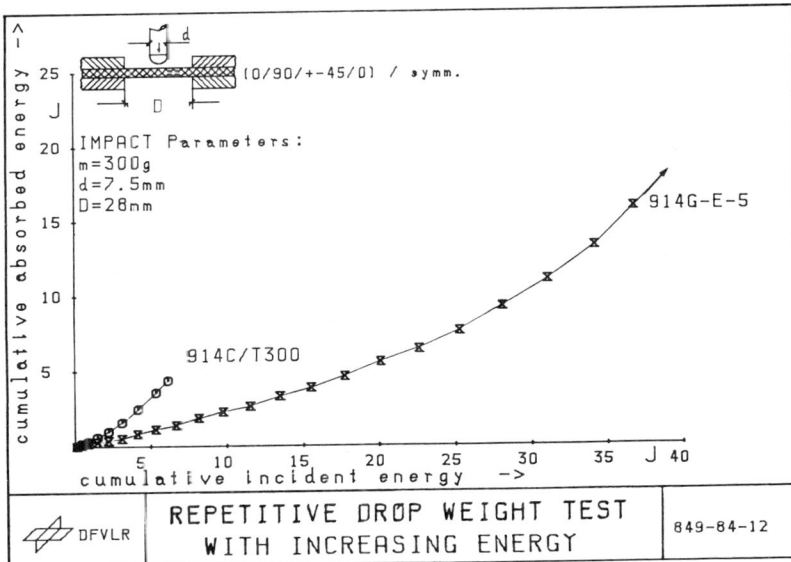

Fig. 6 Comparison of impact resistance of E-glass fibres and T300 carbon fibres with the repetitive drop weight test with increasing energy (RDTIE).

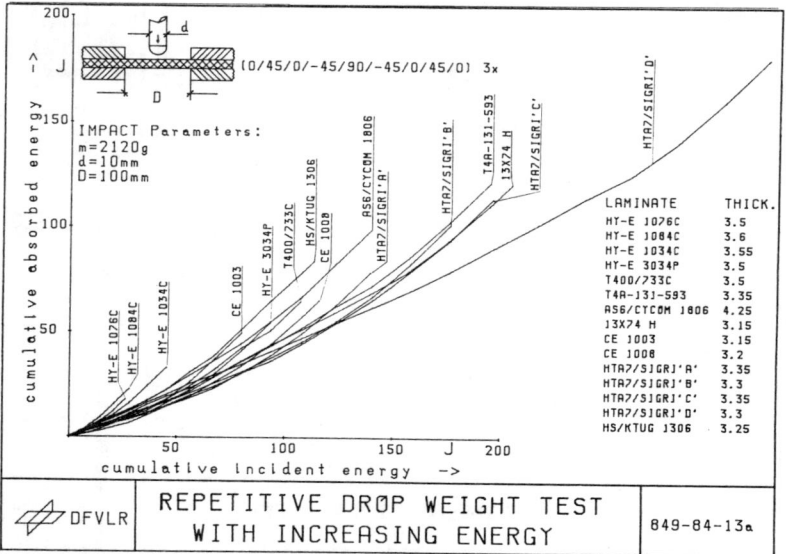

Fig. 7 Comparison of impact properties of various carbon fibre/epoxy systems with the RDTIE.

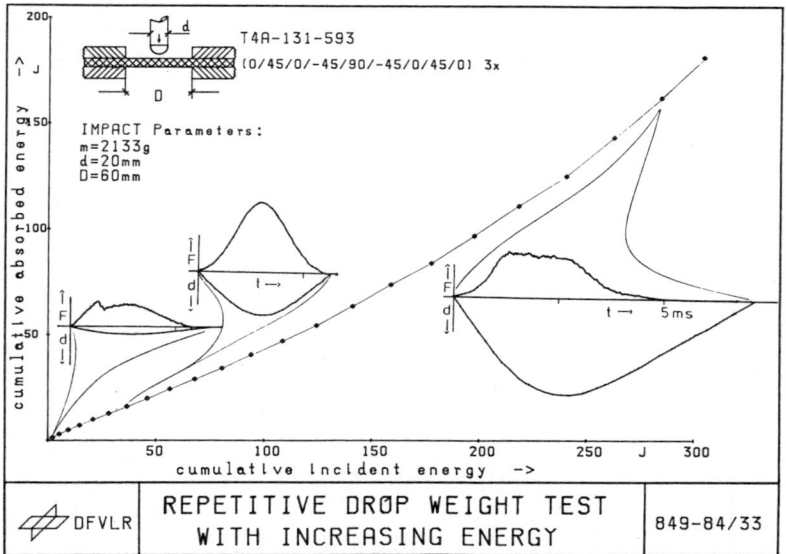

Fig. 8 Slope changes of the RDTIE pseudo curve correspond to damage mode changes.

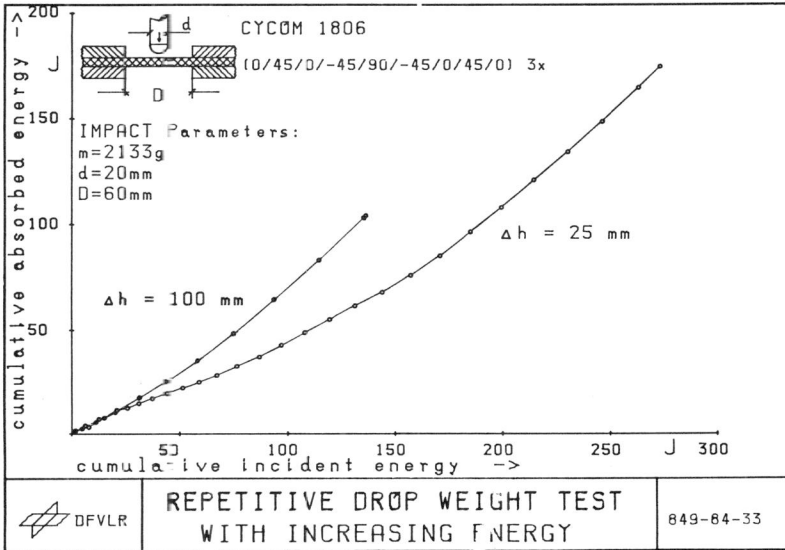

Fig. 9 Influence of impact energy increase rate on RDTIE pseudo curve.

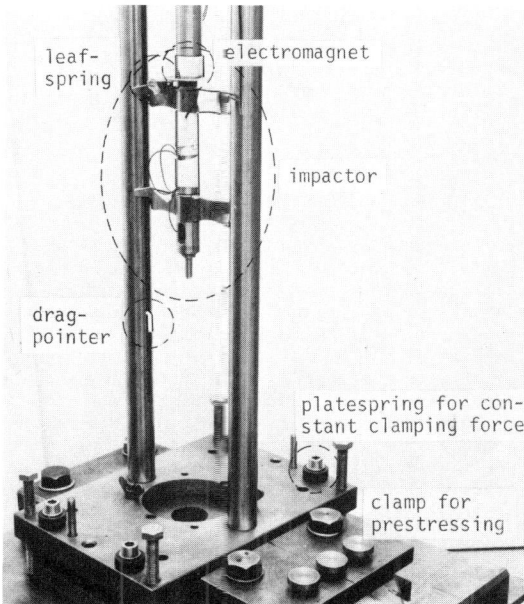

Fig. 10 Drop weight test apparatus with variable clamping diameter, impactor mass and impactor tip and magnetically held drap pointer.

Even with the lowest inpact load cracks in matrix and/or interface parallel
to fibre direction appear. This is clearly proved in the corresponding load-
time curve. After the second or third drop, the slope of the pseudo curve
decreases, reflecting the additional appearance of interlaminar delamination,
proven in a corresponding load-time curve (now the laminate is locally more
compliant, so that first the load can increase only slowly). At a cumulative incident energy of about 140 J the slope of the pseudo curve changes
again, indicating the additional appearance of fibre breaks, reflected in
the corresponding load-time curve by a long period of equalized load level.
The identification of these slope changes makes possible the comparison of
different composites with respect to different damage modes.

The measurement of the incident and absorbed energy does not require any
expensive electronic equipment, e.g. piezo electric transducer. All we
need is a meter stick. The rebounding impactor, with help of a leaf spring,
takes up a light weight drag pointer, which weighs only 0.1 % of the impactor
weight, up to the maximum rebound height (Fig. 10). Then we calculate the
absorbed energy E_{dam}

$$E_{dam} = m * (h_o - h_r)$$

with m = impactor mass, h_o = drop height, h_r = rebound height. Friction
losses can be taken into account approximative.

The repetitive dropweight test with increasing energy at all appearance
is a cheap and reliable method for ranking the impact properties of composites.
But the proper test parameters are not yet defined finally. We use circular
clamping of the specimens in a diameter $D = 15 * t$; t means the laminate
thickness. The hemispherically shaped tip of the impactor has a diameter
d = D/3. The intention of this choice was to make feasible matrix/interface
cracks as far as interlaminar delaminations as far as fibre breaks in a
fairly balanced ratio for a wide range of material behaviour.
The thicknesses of specimens to be compared must be constant. With small
differences, relation to a nominal thickness t_o can be made by

$$E(t_o) = E(t_1) * (t_o/t_1)^2$$

This correction must be done with incident and damage energy.

The influence of the energy increase rate is shown in Fig. 9. It is obvious that the smaller energy increase of about 0.5 J (= 25 mm drop height
increase) allows finer differentiation of the slope of the pseudo curve
than the larger energy increase (= 100 mm drop height increase). The pre-
ceeded diagrams are related to a compromise of 50 mm drop height increase,
which was considered acceptably efficient.
Although undoubtedly some further effort is needed to specify the proper
test parameters, the repetitive dropweight test with increasing energy is
a simple but effective method for testing comparatively the impact behaviour
of composite materials. This test does not give information about the rele-
vance of an impact damage under service conditions ; this must be tested
separately. But the repetitive dropweight test with increasing energy can
be the first step for a proper material selection with respect to impact
resistance.

STRESSES IN PIN-LOADED PLATES

M.W. HYER, D. LIU

Virginia Polytechnic Institute and State University, Blacksburg, VA, U.S.A.
University of Florida, Gainesville, FL, U.S.A.

ABSTRACT

The stresses around the hole in pin-loaded fiber-reinforced glass-epoxy plates are determined experimentally by using transmission photoelasticity. Results are presented for three glass-epoxy plates and an isotropic plate. The latter plate is examined for comparison purposes. The glass-epoxy plates are all 32 layers and have the following lamination sequences : $(0_4/45_4/-45_4/90_4)_s$ $(0)_{32}$, and $(45_4/-45_4)_{2s}$. The geometry of the four plates is the same, the plate-width to hole-diameter ratio being 4 and the end-distance to hole-diameter ratio being 2. The paper presents a brief overview of orthotropic photoelasticity and discusses the method of using the optical data to obtain stresses. The experimental set-up used to load the plates is described. Typical isochromatic and isoclinic fringe patterns are presented. Finally, the radial, circumferential, and friction-induced shear stresses at the hole edge in the three glass-epoxy plates and the isotropic comparison plate are illustrated.
Similarities and differences among the stresses in the four cases are discussed.

INTRODUCTION

The study of stresses in composite bolted joints has often been reduced to the study of the stresses around the hole in a pin-loaded orthotropic plate (1-19). Much of the work has been theoretical in nature, the studies focusing on the effect of plate elastic properties, plate-width to hole-diameter ratios, pin/hole clearance, pin/hole friction, pin elasticity and other parameters on the stresses and the failure load. The experimental work that has been done has involved determining the failure load and failure mode. Some studies have examined the strain levels at a few discrete locations on the plate. While this type of work, particularly the former, is valuable, very little work has been done to examine the overall stress state in a pin-loaded plate. An overall picture of the stress state would serve as a basis for calibrating the stress predictions of theoretical work which generally provides whole-field stress information. The work presented here does provide a glimpse of the overall stress state in a pin-loaded plate. The work here presents the stresses at the hole edge of several pin-loaded orthotropic plates. The stresses for an isotropic plate are included for comparison. The results presented here are based on photoelastic measurements. Specifically, the results are based on transmission photoelastic measurements in fiber-reinforced glass-epoxy. This is opposed to using reflection photoelastic techniques in conjunction with coatings on an opague orthotropic material.

The paper begins by representing a brief overview of transmission photo-elasticity as it is applied to transparent orthotropic materials. Then the experimental set-up used to load plates with a pin through the hole is described. Photographs of typical photoelastic fringe patterns follow. Finally, numerical results for the stress around the hole in the four different plates are presented.

OVERVIEW OF ORTHOTROPIC PHOTOELASTICITY

As a beam of polarized light passes through a transparent material, part of the beam travels at one speed through the thickness while the remaining portion of the beam travels at another speed (20). By using the proper optical elements, the two portions of the beam can be made to interfere when they emerge from the material. It is the purpose of a polariscope to provide the incident polarized light beam and to produce interference of the two emerging portions of the beam. For materials used in the study of stresses, the interference is strongly influenced by the applied stress level. The influence of the stress at a particular point is usually interpreted in terms of the number of fringes at the point. The symbol N is commonly used to denote the number of fringes. The value of N can be interpreted in terms of the components of the birefringence tensor, namely

$$N^T = \sqrt{(N_{22}^T - N_{11}^T)^2 + (N_{12}^T)^2} \tag{1}$$

The quantities N_{11}^T, N_{22}^T, and N_{12}^T are the components of the total birefringence tensor as measured in the 1-2 principal material system of the ortho-tropic material. The superscript T is used here to denote that the discussion will involve two types of birefringence, total (T) and residual (R). The total birefringent effect is the sum of the effect due to the applied stresses and the residual effects in the material. Residual effects are those associated with the fabrication of the material and are present with no applied stress. The relationship among the various birefringence effects is given by

$$\begin{Bmatrix} N_{11}^T \\ N_{22}^T \\ N_{12}^T \end{Bmatrix} = \begin{vmatrix} a_{11} & a_{11} & 0 \\ a_{21} & a_{22} & 0 \\ 0 & 0 & 0 \end{vmatrix} \begin{Bmatrix} \sigma_1 \\ \sigma_2 \\ \tau_{12} \end{Bmatrix} + \begin{Bmatrix} N_{11}^R \\ N_{22}^R \\ N_{12}^R \end{Bmatrix} \tag{2}$$

where σ_1, σ_2, and τ_{12} are the stresses referred to the principal material system. The a_{ij}'s are the stress-birefringence coefficients of the material and the relation is assumed to be orthotropic in the principal material system.

A common polariscope can measure the magnitude of the birefringence, N^T, but not the individual components, N^T_{11}, N^T_{22} and N^T_{12}. A polariscope also measures the total isocline angle θ^T. The isocline angle identifies the principal directions of the birefringence tensor. To utilize the capability of a polariscope, the above equation is put in the form

$$\frac{\sigma_1}{f_1} - \frac{\sigma_2}{f_2} = N^T \cos(2\theta^T) - N^R \cos(2\theta^R) \tag{3a}$$

and

$$\frac{\tau_{12}}{f_{12}} = \frac{N^T}{2} \sin(2\theta^T) - \frac{N^R}{2} \sin(2\theta^R) \tag{3b}$$

N^R is defined similar to N^T in eq. 1 and it is the magnitude of the bire-fringence observed when no stresses are applied. θ^R is the isocline observed when no stresses are applied. It is sometimes referred to as the residual isocline while N^R is referred to as the residual fringe.
In the above equations both θ^T and θ^T are measured relative to the 1-axis of the principal material system. The quantities f_1, f_2, and f_{12} are defined by

$$f_1 = \frac{1}{a_{11} - a_{21}} \qquad f_2 = \frac{1}{a_{22} - a_{12}} \qquad f_{12} = \frac{1}{a_{66}} \tag{4}$$

Values for f_1, f_2, f_{12}, N^R and θ^R are obtained for a particular photoelastic material by calibration. Equations 3a and 3b are referred to as the stress-optic equations.

In an x-y or r-θ system oriented at an angle ϕ relative to the 1-2 system, the stress optic equations become, after transformation,

$$\frac{\sigma_x}{C_1} - \frac{\sigma_y}{C_2} + \frac{\tau_{xy}}{C_3} = N^T \cos(2\theta^T) - N^R \cos(2\theta^R) \tag{5a}$$

$$\frac{\sigma_x - \sigma_y}{C_4} + \frac{\tau_{xy}}{C_5} = \frac{N^T}{2} \sin(2\theta^T) - \frac{N^R}{2} \sin(2\theta^R) \tag{5b}$$

with

$$\frac{1}{C_1} = (\frac{\cos^2\phi}{f_1} - \frac{\sin^2\phi}{f_2}) \qquad \frac{1}{C_2} = (\frac{\cos^2\phi}{f_2} - \frac{\sin^2\phi}{f_1})$$

$$\frac{1}{C_3} = 2(\frac{1}{f_1} + \frac{1}{f_2}) \sin\phi \cos\phi \qquad \frac{1}{C_4} = -\frac{\sin\phi \cos\phi}{f_{12}} \tag{6}$$

$$\frac{1}{C_5} = \frac{\cos^2\phi - \sin^2\phi}{f_{12}} \quad .$$

This is the form of the photoelastic equations used here.

As can be seen from eqs. 3 and 5, at a point there are three stresses to be determined and only two experimental measures that are related to these three stresses, namely value of N^T and θ^T. A third relation involving the stresses is needed to uniquely determine all three components of stress. In the present work the third relation among stresses was provided by the plane-stress form of the equilibrium equations. Specifically, the equations were used in finite-difference form, a finite-difference mesh being superposed over the region of the plate where the stresses were to be determined.

Since there are two equilibrium equations for plane stress, enforcing both equations in addition to enforcing the two stress-optic equations, eqs. 3 or 5, leads to four equations for three stresses at a point. In addition, there are certain boundary conditions the stresses must satisfy. For example the normal and shear stresses are zero along any traction-free boundary of the plate. Considering all finite-difference points simultaneously (the finite-difference kernels couple the points), there results an overdetermined set of linear algebraic equations for the three stresses at each point in the mesh. The solution of this overdetermined set is a solution for the stresses that satisfy all equations in a least-squares sense (21). All numerical results presented here were determined in this manner.

EXPERIMENTAL SET-UP

The material used in this study consisted of approximately 55 % volume fraction glass fiber in an epoxy matrix (22). The indicies of refraction of the glass fibers and the epoxy were closely matched so the material was quite transparent. Three glass-epoxy plates and one comparison isotropic plate were tested. The glass-epoxy plates were all 32 plies in construction. The three glass-epoxy plates were : quasi-isotropic, $(0_4/45_4/-45_4/90)_s$; unidirection, $(0)_{32}$; and angle-ply, $(45_4/45_4)_{2s}$. The ply orientations are relative to the load direction.

Figure 1 illustrates a plate in the fixture used to transmit a load to the hole. Dead weights were used to provide the force P and the loading fixture was situated between the two halves of a split-bench polariscope. The two long aluminium plates were used so that the stresses in the glass-epoxy plate, bore down on a steel pin which protruded from the hole. The crossbars were connected to a U-shaped yoke which transmitted the load to the lower portion of the loading frame. The yoke arrangement allowed for a full field of view of the region of high stress, namely the region below the net section. Since this was the primary area of interest, no optical data were gathered above the steel crossbars. Steel pins 9.5 mm in diameter were used to connect the components of the loading fixture together. The plates were 2.29 thick and the plate speciments were cut to be 203 mm wide and 292 mm long. The hole for the pin was 50.8 mm in diameter. With this geometry, the plate-width to hole-diameter ratio, W/D, was 4. The distance from the center of the hole to be unloaded or free end of the plate, e, was 101 mm. Hence e/D was equal to 2. This geometry was the only one tested. The isotropic plate 3.00 mm thick. The hole in each plate was drilled with an ultrasonic core drill to minimize residual birefringence effects and damage at the hole.

Table 1 indicates the elastic and optical properties of the plates tested. The properties are referred to the x-y coordinate system of Fig. 1.

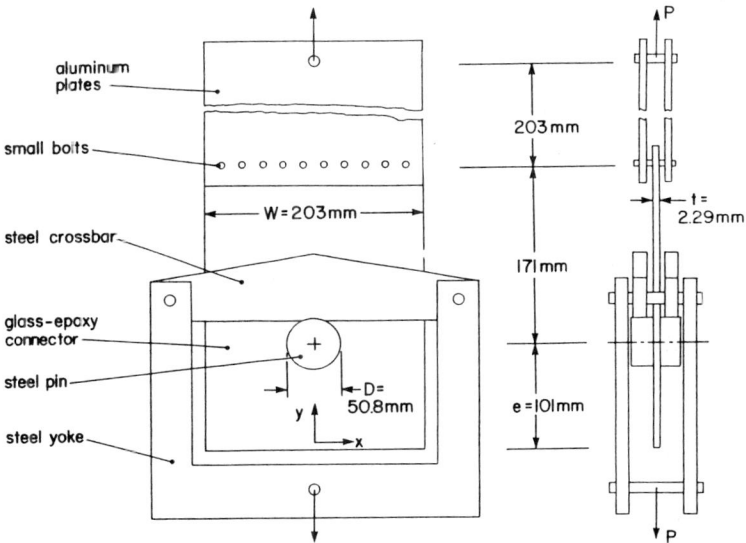

Fig. 1 Plate geometry and loading fixture.

Table 1 Mechanical and optical properties of specimens tested.

Plate property	Isotropic	Quasi-Isotropic	Unidirectional	Angle-ply
E_x, GPa	1.38	19.7	37.2	12.4
E_y, GPa	1.38	19.7	12.3	12.4
G_{xy}, GPa	0.49	7.38	3.93	10.9
ν_{xy}	0.4	0.328	0.300	0.577
f_x, kPa/fringe/m	≤1.0	96.8	129	674
f_y, kPa/fringe/m	≤1.0	96.8	68	674
f_{xy}, kPa/fringe/m	≤1.0	96.8	57	115
N^R, fringe	0	0	0.12	0
θ^R,	0	0	0	0

Figure 2 illustrates the lightfield isochromatic fringe pattern in the $(0)_{32}$ plate. The fibers, which are aligned with the load direction, can be seen in the fringes. The load (P of Fig. 1) is 5.92 kN. The fringes shown in the figure are typical of the fringe patterns in fiber-reinforced materials. The fringes are not as sharp and distinct as in homogeneous isotropic materials. The scattering of the light by the fibers is the primary cause. However, imperfections in the form of bubbles or small opaque regions due to poor fiber-matrix or interlaminar bonding have some influence. From the figure it is also evident that there are not a large number of fringes generated. This is due to the insensitivity of the glass-epoxy material to the birefringent effect. The application of more load would remedy the situation. However, applying more load would increase the risk of crazing the material in regions of high stress. This causes the material to become opaque and so photoelastic information is not available in that region. Obviously it is exactly these regions of high stress that are of paramount interest. Finally, it is evident there is fair symmetry, about the vertical centerline, of the fringe pattern. The lack of perfect symmetry is not an indication of load asymmetry. Lack of spatially uniform material properties, particularly optical properties, is the primary cause. The fringe patterns from the more sensitive and extremely uniform (spatially) isotropic material plate indicated a high degree of symmetry with the loading fixture. Thus it is clear that the stress-optic coefficients in Table 1 represent average properties.

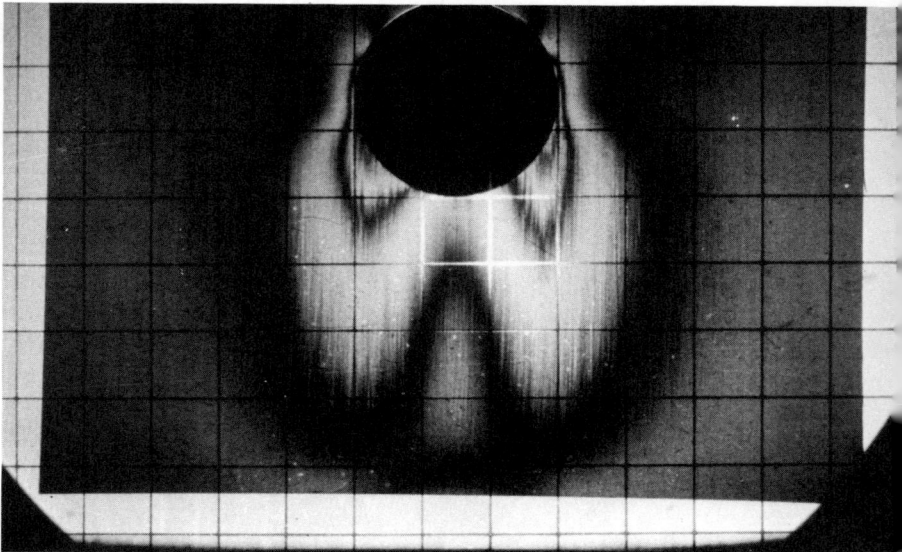

Fig. 2 Lightfield isochromatic fringes of the unidirectional plate, load = 5.92 kN.

Figure 3 shows a typical isoclinic fringe pattern. This is the 30° isocline (T = 30°) in the quasi-isotropic plate. Because these fringes were
poorly defined, high-contrast film was used to clarify them. (Fig. 3 used
high-contrast film).

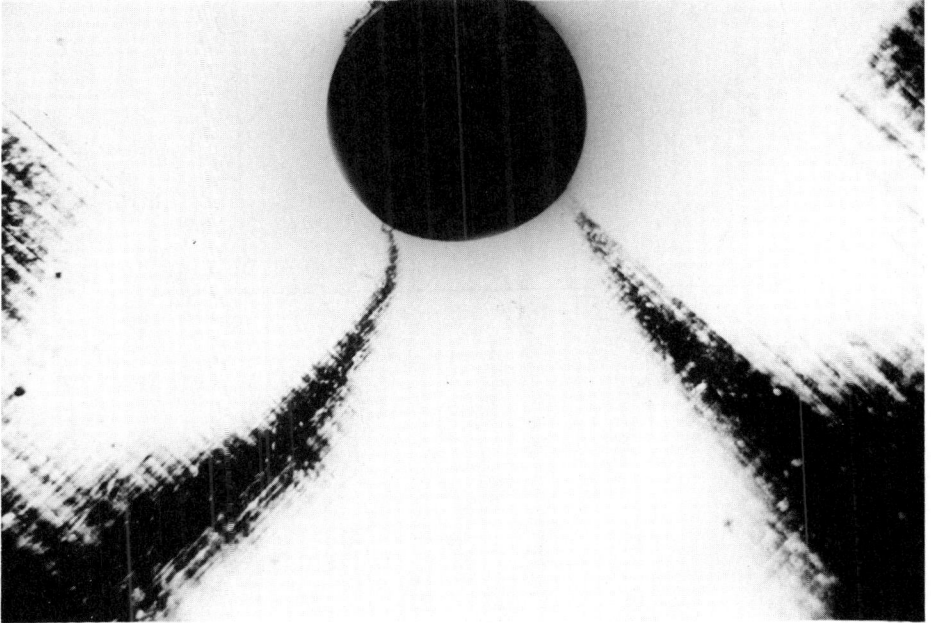

Fig. 3 Thirty-degree isoclinic fringe of the quasi-isotropic plate.

EXPERIMENTAL RESULTS : CALCULATES STRESSES

Figures 4-7 show the hole-edge stresses for the four plates. The stresses
shown are the radial (bearing) σ_r, circumferential σ_θ, and the friction-
induced shear $\tau_{r\theta}$. These stresses are shown as a function of circumferential
location around the hole, 0° being the bottom of the hole and 90° being the
net section. The stresses are normalized by the average bearing stress S.
The quantity S is defined as

$$S = \frac{P}{Dt}$$

where P is the applied load, t is the plate thickness, and D is the hole
diameter. Superposed on the figures is the cosinusoidal stress distribution
often used in analytical studies.

Fig. 4 Hole-edge stresses in the isotropic plate.

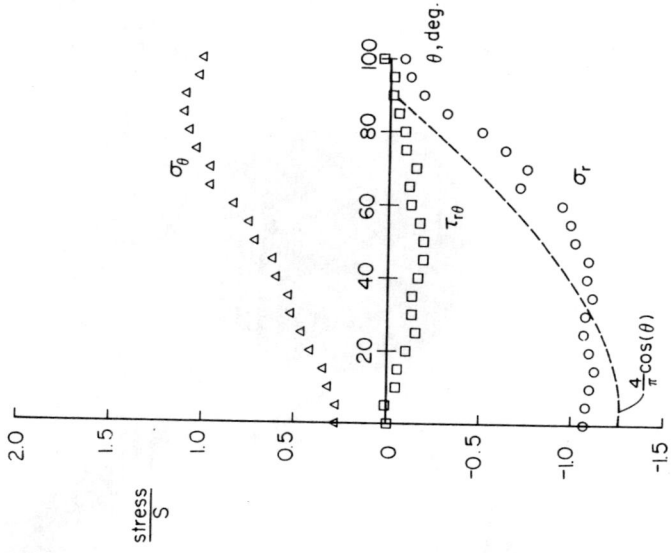

Fig. 5 Hole-edge stresses in the quasi-isotropic plate

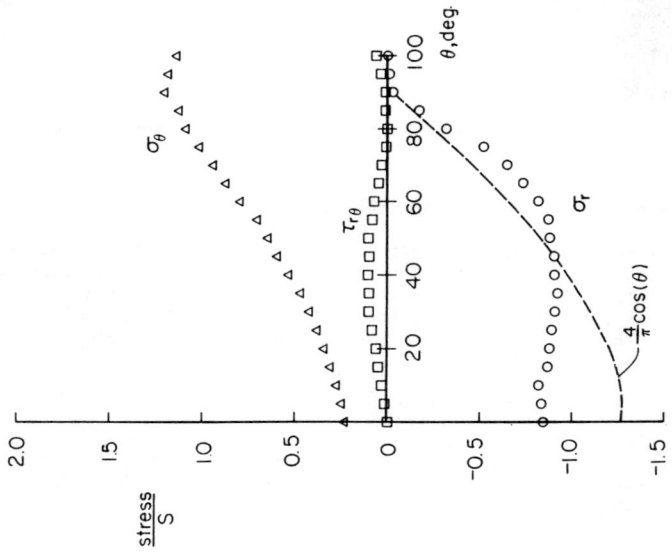

Fig. 6 Hole-edge stresses in the unidirectional plate.

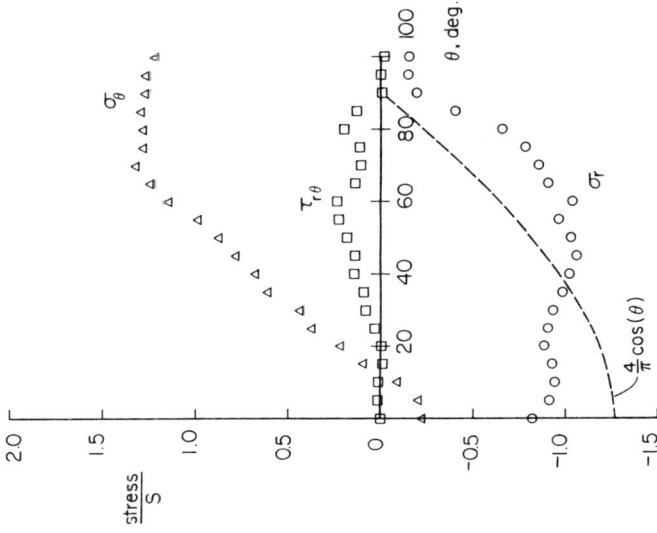

Fig. 7 Hole-edge stresses in the angle-ply plate.

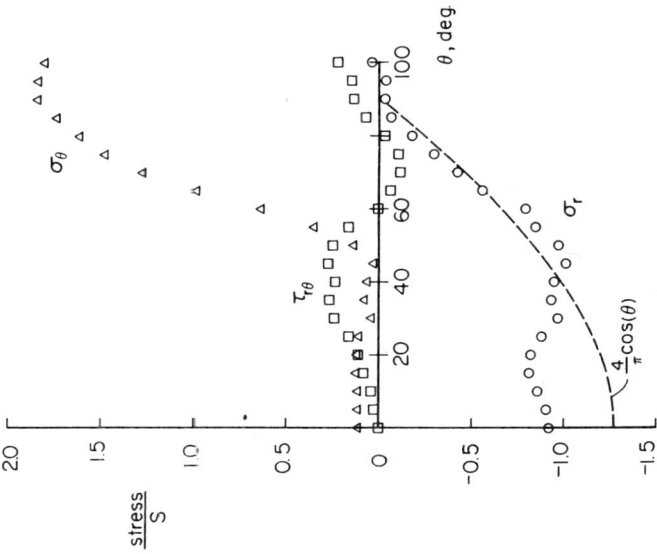

Several characteristics of the stress states in the plates are evident from the figures. First, the cosinusoidal distribution does not accurately represent the radial stress in the range $0° < \theta < 30°$. The actual radial stress in this range is less than the cosinusoidal representation. This particular inaccuracy in the cosinusoidal representation had been shown to be due to frictional effects (13,19). Second, the radial stress distribution for all four cases are, generally speaking, quite similar. Both the magnitudes and the distributions with circumferential position are similar. It was expected that this would not be the case. It was expected that differences in the degree of orthotropy from one laminate to the next would produce more striking contrasts. Apparently the 3:1 ratio of stiffnesses (i.e. E_x/E_y) for the unidirectional case was not enough different than the 1:1 ratio of the other three cases. Third, the circumferential stress distributions are different, there being the most severe net-section stress concentration for the unidirectional case. For this plate, in the range $0° \leqslant \theta \leqslant 50°$ the circumferential stresses are small. For $\theta > 50°$, the stresses gradually increase as θ increases from $0°$ to $90°$. In the angle-ply plate the circumferential stress is actually compressive at $\theta = 0°$. Such an effect can be caused by friction. Finally, for reasons as yet not explained, the sign of the shear stress for the quasi-isotropic plate was opposite in sign to the shear stresses in the other plates.

Global equilibrium checks, made by numerically intergrating the stresses around the hole-edge, indicated the stresses computed to be consistent. The integral, which involved the radial and shear stresses, was within 10 % of the value of the applied load.

A further discussion of the stress state in these plates is available in ref. 23. More details of the scheme to determine the stresses from the optical data is also available in this reference.

ACKNOWLEDGEMENTS

The research effort which led to the results presented here was financially supported by the United States Army Research and Technology Laboratories (Aviation Systems Command). The technical monitor was Donald J. Baker of the National Aeronautics and Space Administration's Langley Research Center.

REFERENCES

1. Waszczak, J.P. and Cruse, T.A., "Failure Mode and Strength Predictions of Anisotropic Bolt Bearing Specimens", J. Comp. Mat., vol. 5, July 1971, pp. 421-425.

2. Oplinger, D.W. and Gandhi, K.R., "Stresses in Mechanically Fastened Orthctropic Laminates", 2nd. Conf. of Fib. Comp. in Flight Vehicle Design, May 1974, pp. 813-841.

3. Oplinger, D.W. and Gandhi, K.R., "Analytical Studies of Structural Performance in Mechanically Fastened Fiber-Reinforced Plates", Army Symp. on Solid Mech., Sept. 1974, pp. 211-240.

4. De Jong, T., "Stresses Around Pin-Loaded Holes in Elastically Orthotropic or Isotropic Plates", J. Comp. Mat., vol. 11, July 1977, pp. 313-331.

5. Wilkinson, T.L., "Stresses in the Neighborhood of Loaded Holes in Wood and Applications to Bolted Joints", Ph.D. Dissertation, Univ. of Wisc.-Madison, Available through Univ. Microfilms, Ann Arbor, MI, 1978.

6. Agarwal, B.L., "Static Strength Prediction of Bolted Joint in Composite Materials", AIAA Journal, vol. 18 N. 11, Nov. 1980, pp. 1371-1375.

7. Mangalgiri, P.D. a,d Dattaguru, B., "Elastic Analysis of Pin Joints in Composite Plate", Rep. nO. ARDB-STR-5014, Dept. Aeronaut. Engr., Ind. Inst. Sci., Nov. 1980.

8. Wong, C.M.S. and Matthews, F.L., "A Finite Element Analysis of Single and Two-Hole Bolted Joints in Fibre-Reinforced Plastic", J. Comp. Mat., vol. 15, Sept. 1981, pp. 481-491.

9. De Jong, T. and Vuil, H.A., "Stresses Around Pin-Loaded Holes in Elastically Orthotropic Plates with Arbitrary Load Direction", Report LR-333, Dept. of Aerospace Engr., Delft Univ. Tech., Sept. 1981.

10. Crews, J.H. and Hong, C.S., "Stress-Concentration Factors for Finite Orthotropic Laminates with a Pin-Loaded Hole", NASA TP 182, May 1981.

11. Garbo, S.P. and Ogonowski, J.M., "Effects of Variances and Manufacturing Tolerances on the Design Strength on Life of Mechanically Fastened Composite Joints', AFWAL-TR-81-3041, vols. 1-3, April 1981.

12. Soni, S.R., "Failure Analysis of Composite Laminates with a Fastener Hole", ASTM STP 749, Joining of Composite Materials, 1981, p. 145-165.

13. De Jong, T., "The Influence of Friction on the Theoretical Strength of Pin-Loaded Holes in Orthotropic Plates", Rep. No. LR-350, Dept. of Aerospace Engr., Delft Univ. Tech., March 1982.

14. Matthews, F.L., Wong, C.M. and Chryssafitis, S., "Stress Distribution Around a Single Bolt in Fibre-Reinforced Plastic", Composites, vol. 13, No. 3, July 1982, pp. 316-322.

15. Rowlands, R.E., Rahman, M.U., Wilkinson, T.L. and Chiang, Y.I., "Single- and Multiple-Bolted Joints in Orthotropic Materials", Composites, vol. 13, No. 3, July 1982, pp. 273-278.

16. Chang, F.K., Scott, R.A. and Springer, G.S., "Strength of Mechanically Fastened Composite Joints", J. Comp. Mat., vol. 16, Nov. 1982, pp. 470-494.

17. Chang, F.K., Scott, R.A. and Springer, G.S., "Failure of Composite Laminates Containing Pin-Loaded Holes - Method of Solution", J. Comp. Mat., vol. 18, May 1984, pp. 255-278.

18. Chang, F.K., Scott, R.A. and Springer, G.S., "Design of Composite Laminates Containing Pin-Loaded Holes", J. Comp. Mat., vol. 18, May 1984, pp. 279-289.

19. Hyer, M.W. and Klang, E.C., "Stresses in Pin-Loaded Orthotropic Plates". Virginia Polytechnic Institute and State University, Center for Composite Materials and Structures, Report No. CCMS-84-02, 1984.

20. Dally, J.W. and Riley, W.F., "Experimental Stress Analysis", 2nd edition, McGraw-Hill, New York, 1978, Ch. 13.

21. Berghaus, D.G., "Overdetermined Photoelastic Solution Using Least-Squares", Experimental Mechanics, vol. 13, no. 3, pp. 97-204, 1973.

22. Daniel, I.M., Niro, T. and Koller, G.M., "Development of Orthotropic Birefringent Materials for Photoelastic Stress Analysis", NASA CR-165709, May 1981.

23. Hyer, M.W. and Liu, D., "Stresses in Pin-Loaded Orthotropic Plates Using Photoelasticity", NASA-CR (number not yet assigned), 1985.

DIRECT SHEAR COMPLIANCE MEASUREMENT FOR FIBRE REINFORCED COMPOSITES

D. Van Gemert

Catholic University of Leuven, Belgium

INTRODUCTION

Many applications of composite materials in civil and mechanical engineering involve thin laminate structures, loaded in a state of plane stress. The laminated plate theory generally used in these cases to estimate strengths and rigidities requires the static, unidirectional, mechanical properties of each constituent layer.

Composite materials mostly have low shear strengths and moduli relative to their longitudinal properties. It is therefore necessary to determine the in-plane shear stress/strain curve with a fair accuracy over a pratically useful shear strain range.

The fabrication of test specimens consisting of only one layer is often impossible. In such cases it is necessary to conduct tests on multilayer specimens and use lamination theory to calculate the single-layer properties. Since there is no simple test method to induce a state of pure shear in a specimen, there are numerous approximate test methods which can be used.

In this we will discuss the results, obtained with :

- a panel shear test, in which a thin, square specimen is loaded in shear along each side
- a rail shear test, in which opposite sides of a thin, rectangular specimen are attached to loading rails and the remaining sides are free.

The tested laminates are glass-fibre reinforced polyester plates, supplied by Polyship N.V. at Oostende. The fibre reinforcement consists of mats of woven strands. The GRP is used as non-magnetic material for army ships. The laminates are manufactured by mat laying in a hand lay-up procedure. The laminate consists of five mat-layers. The total thickness if 5 mm. Mean volume fractions are :

$$v_f = 0,32 \qquad v_m = 0,66 \qquad v_v = 0,02 \qquad (1)$$

Although is not really the case, we assume that we can analyse the laminate as a composition of UD-laminae. So we neglect the influence of curvatures of the fibres due to woving.
With a simple rule-of-mixtures (ref. [1]) the theoretical elastic moduli of the fictitious corresponding unidirectional lamina were calculated

$$E_{11} = 25048 \text{ MPa} \qquad \nu_{12} = 0,23$$

$$E_{22} = 6226 \text{ MPa} \qquad \nu_{21} = 0,06 \qquad (2)$$

$$G_{12} = 2399 \text{ MPa}$$

The equations, relating elastic moduli of laminae and laminates, are listed in the Appendix. With these equations the moduli of the lamina can be calculated from the experimentally determined values of the elastic constants of the laminate.

Experiments for shear are inteded to apply a shear force T_{12} to the specimen, and to measure γ_{12}. From the compliance S_{66} we can calculate the shear modulus G_{12}.

PANEL SHEAR TEST

Panel shear specimens are tested in a four link rig shown schematically in Fig. 1. The specimen size is 150 mm x 150 mm, from which the central square of 70 mm x 70 mm is the measuring area, subjected to pure shear, Fig. 2.

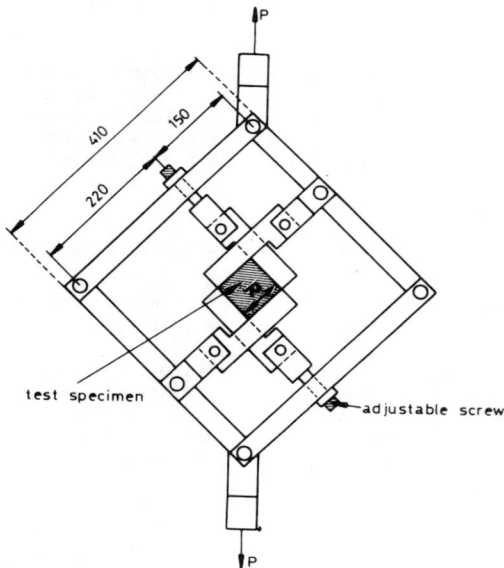

Fig. 1 Panel shear test rig.

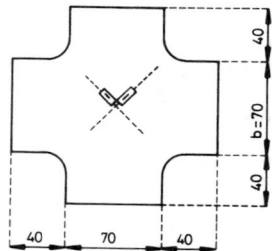

Fig. 2 Panel shear specimen

The shear force on each side of the specimens is equal to

$$P \cdot \frac{L}{b} \cdot \cos 45 \tag{3}$$

The shear stress is calculated as

$$\tau_{12} = \frac{1}{bt} \cdot P \frac{L}{b} \cdot \cos 45 \tag{4}$$

where t is the thickness of the specimen and b the side length of the test specimen. L is the length of the rig beams.
Straingauges are placed in the 45°-direction (Fig. 2).
From Weyrauch's circle for strains we get

$$\gamma_{12} = 2 \, \varepsilon_{45} \tag{5}$$

As a $(0/90)_s$ laminate is orthotropic, its constitutive relation is expressed as

$$\begin{Bmatrix} \varepsilon_1 \\ \varepsilon_2 \\ \gamma_{12} \end{Bmatrix} = \begin{vmatrix} S_{11} & S_{12} & 0 \\ S_{21} & S_{22} & 0 \\ 0 & 0 & S_{66} \end{vmatrix} \begin{Bmatrix} N_1 \\ N_2 \\ T_{12} \end{Bmatrix} \tag{6}$$

If only a shear load T_{12} is applied the shear stiffness and compliance can easily be calculated

$$\frac{1}{S_{66}} = A_{66} = \frac{T_{12}}{\gamma_{12}} = \frac{P \ L \ \cos 45}{2b^2 \ \varepsilon_{45}} \tag{7}$$

The shear modulus is now derived from

$$S_{66} = \frac{1}{G_{12} \cdot t} \tag{8}$$

or

$$G_{12} = \frac{1}{S_{66} \cdot t} = \frac{P \ L \ \cos 45}{2b^2 \ t \ \varepsilon_{45}} \tag{9}$$

In executing the test great care must be taken to keep the test rig exactly in an orthogonal shape. This can be done by means of the adjustable screws. Fixing of the side holders to the specimen is done by combined bolting and fluing, to get a uniform transfer of forces. Even with these precautions the reproducibility of the test is problematic. For six tests we found $G_{12} = 1877$ MPa with standard deviation equal to 135 MPa.

RAIL SHEAR TEST

In the rail shear test the specimen is loaded in pure shear as is schematically indicated in Fig. 3.

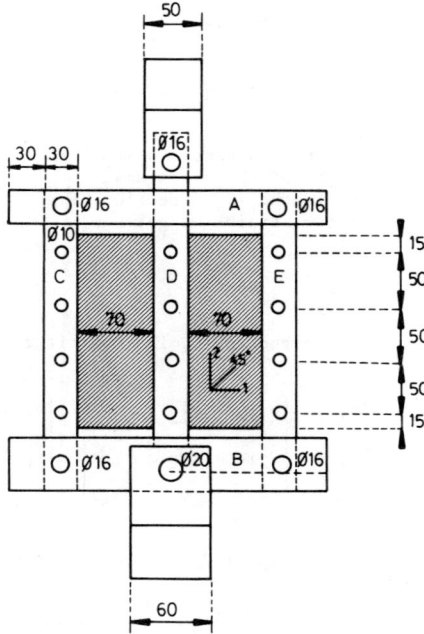

Fig. 3   Rail shear test set-up

From the symmetry of the structure and the loading system we get for the stresses and strains

$$\tau_{12} = \frac{P}{2A} \qquad \gamma_{12} = 2\,\varepsilon_{45} \qquad (10)$$

where A is the cross-sectional area of the specimen parallel to the load.

Analogously as in the panel shear test we find

$$\frac{1}{S_{66}} = A_{66} = \frac{\tau_{12}}{\gamma_{12}} = \frac{P}{4b \cdot \varepsilon_{45}} \qquad (11)$$

from which we calculate G_{12} as

$$G_{12} = \frac{P}{4b \ t \ \varepsilon_{45}} \tag{12}$$

The strains are measured with strain gauges, placed at an angle of 45° to the applied load. The test set-up must be very accurate, and transfer of force from the rails to the specimen is often non-uniform, especially if only a bolted connection is used.
Therefore we combined the bolted connection with an epoxy bonding over the toal length of the rails. This procedure gives good results, but it makes the test preparation very laborious.
A typical stress-strain curve is shown in Fig. 4.

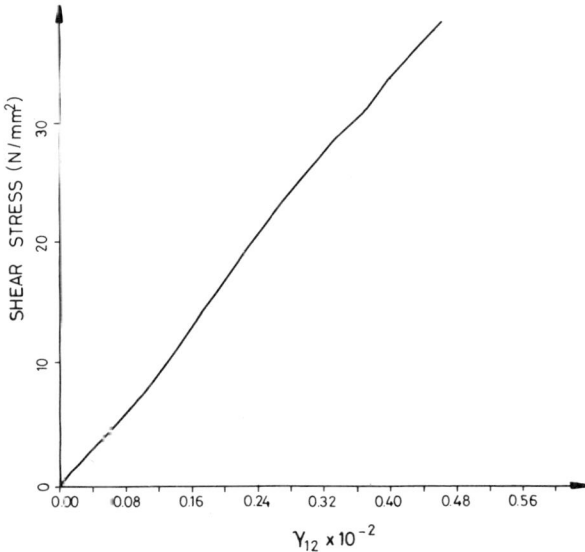

Fig. 4   Typical stress-strain curve for rail shear test

The calculation of the shear modulus is done by means of a linear regression procedure. The reproducibility of the rail shear test is quite satisfactory.
For six tests we found a mean value of 1657 MPa for the shear modulus G_{12}, with a standard deviation of 58 MPa. The stress level was limited to 8 MPa.

TENSILE SHEAR TEST

For comparison we executed tensile tests on linear specimens in ± 45° off-axis configuration, Fig. 5.

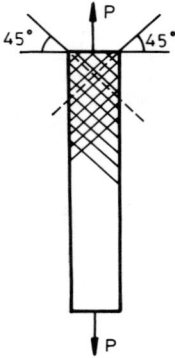

Fig. 5 45° off-axis tensile test

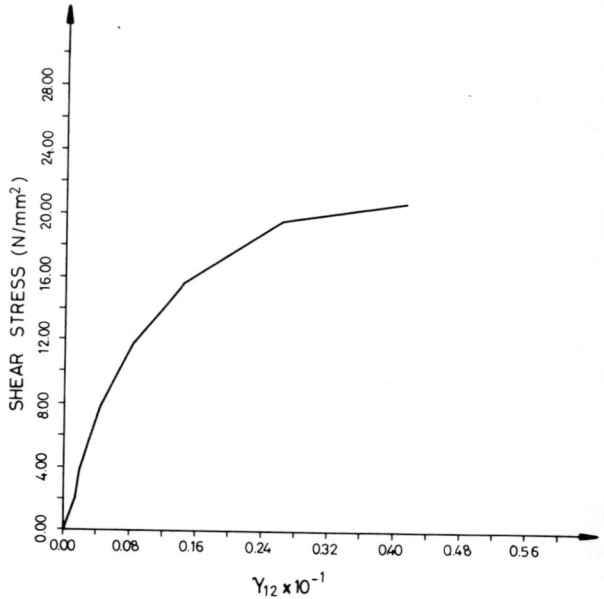

Fig. 6 Shear stress-strain curve in off-axis tensile test.

Strain gauge measurements are done in the longitudinal and transversal directions. For this loading case the shear stresses and strains are calculated as

$$\tau_{12} = \frac{P}{2A} \qquad \gamma_{12} = \varepsilon_L - \varepsilon_D \qquad (13)$$

where A is the cross-sectional area of the specimen.
A typical stress-strain curve is shown in Fig. 6. By linear regression for low stress levels we found

$$G_{12} = 1488 \text{ MPa} \quad (s = 174 \text{ MPa}).$$

CONCLUSION

The results obtained for the shear modulus of a glass fibre reinforced polyester by means of three different test methods are compared in Table 1. The value, theoretically calculated from the rule of mixture, is also given.

Test method	G_{12} (MPa)	s (MPa)
Panel shear	1877	135
Rail shear	1657	58
Off-axis 45°	1488	174
Theoretical	2399	

Table 1   Experimental results.

The rail shear test and the off-axis tensile test are in "reasonable" agreement, although compared to the theoretical value it seems to be very optimistic to speak about agreement.   The results show very clearly that the determination of the shear characterstics of composites is a difficult and complex task, and that all experimental data must be handled with great cautiousness.   Having in mind the difficulties, encountered in preparing and executing a panel shear or a rail shar test, as well as the accuracy and agreement which is obtained, it seems to be realistic to limit ourselves to simple and easy to execute off-axis tensile tests.

REFERENCES

1.   Tsai, S.W., Hahn, N.T., Introduction to Composite Materials. Technomic Publ. Co, 1980

2.   Van Gemert, D., Moyson, E., Inleiding tot de laminaattheorie. Proceedings "Composite Materials Workshop", Leuven 4-6 juni, 1984.

3.   Yeow, Y.T., Brinson, H.F., A comparison of simple shear characterization methods for composite laminates. Composites, pp. 49-55, 1978.

4.   Terry, G., A comparative investigation of some methods of unidirectional, in-plane shear characterisation of composite materials. Composites, pp. 233-237, 1979.

APPENDIX : STIFFNESS AND COMPLIANCE MATRICES

- On-axis compliance matrix $\underline{S}$ for UD

$$\begin{Bmatrix} \varepsilon_x \\ \varepsilon_y \\ \gamma_{xy} \end{Bmatrix} = \begin{vmatrix} \dfrac{1}{E_x} & -\dfrac{\nu_y}{E_y} & 0 \\ -\dfrac{\nu_x}{E_x} & E_y & 0 \\ 0 & 0 & \dfrac{1}{G} \end{vmatrix} \begin{Bmatrix} \sigma_x \\ \sigma_y \\ \tau_{xy} \end{Bmatrix} \quad \text{or} \quad \underline{\varepsilon} = \underline{S} \cdot \underline{\sigma}$$

- On-axis stiffness matrix $\underline{Q}$ for UD lamina

$$\begin{Bmatrix} \sigma_x \\ \sigma_y \\ \tau_{xy} \end{Bmatrix} = \begin{vmatrix} \dfrac{E_x}{1-\nu_x \nu_y} & \dfrac{\nu_y E_x}{1-\nu_x \nu_y} & 0 \\ \dfrac{\nu_x E_y}{1-\nu_x \nu_y} & \dfrac{E_y}{1-\nu_x \nu_y} & 0 \\ 0 & 0 & G \end{vmatrix} \begin{Bmatrix} \varepsilon_x \\ \varepsilon_y \\ \gamma_{xy} \end{Bmatrix} \quad \text{or} \quad \underline{\sigma} = \underline{Q} \cdot \underline{\varepsilon}$$

- Off-axis compliance matrix S and stiffness matrix Q are calculated by means of transformation matrices from off-axis to on-axis elements. The elaborated transformation formulas are given in ref. [2].

- In symmetrical laminates the following relations hold

$$\begin{Bmatrix} N_1 \\ N_2 \\ T_{12} \end{Bmatrix} = \begin{vmatrix} A_{11} & A_{12} & A_{16} \\ A_{21} & A_{22} & A_{26} \\ A_{61} & A_{62} & A_{66} \end{vmatrix} \begin{Bmatrix} \varepsilon_1^o \\ \varepsilon_2^o \\ \gamma_{12} \end{Bmatrix}$$

where $\quad A_{ij} = \displaystyle\int_{-\frac{1}{2}h}^{+\frac{1}{2}h} Q_{ij} \, dz$

In orthotropic laminates $A_{16} = A_{26} = A_{61} = A_{62} = 0$.

SOME ANALYTICAL SOLUTIONS FROM AN APPROXIMATE CONTINUUM THEORY FOR ANISOTROPIC COMPOSITES

J.L. KING

University of Edingourgh, Scotland

SUMMARY

We consider two-dimensional stress problems for a composite material consisting of linear-elastic fibres and matrix. For an anisotropic continuum, the stress function obeys a modified biharmonic equation and few exact solutions exist. With some approximations, this can be reduced to Laplace's equation, permitting discussion of two important problem areas, the flexure of beams and the stresses at notch roots. Both of these areas have also been investigated using the theory for an "ideal" composite ; a comparison is made of the results derived from the two theories.

OUTLINE OF THEORY

The theoretical model on which the present work is based is given in greater detail in [1], where there is some analysis of the assumptions made and the validity of the approximations involved.
Here we give a summary of the theory.

For an anisotropic continuum, the governing equation for the stress function, χ, is

$$(\frac{\partial^2}{\partial x^2} + \beta_1^2 \frac{\partial^2}{\partial y^2}) (\frac{\partial^2}{\partial x^2} + \beta_2^2 \frac{\partial^2}{\partial y^2}) \chi = 0 \tag{1}$$

With fibres lying in the x-direction,

$$\beta_1^2 = \frac{E_y}{G_{xy}} = 0 \quad (1) \quad , \quad \beta_2^2 = \frac{G_{xy}}{E_x} \ll 1 \quad ,$$

E_x and E_y being the two Young's moduli and G_{xy} the shear modulus. (For carbon fibre, some experimental results give $\beta_1 = 1.13$, $\beta_2 = 0.16$).

With direct stress components σ_x, σ_y, (1) becomes

$$(\frac{\partial^2}{\partial x^2} + \beta_2^2 \frac{\partial^2}{\partial y^2}) (\beta_1^2 \sigma_x + \sigma_y) = 0$$

In typical situations, $\sigma_x \gg \sigma_y$ suggesting the controlling equation

$$\left(\frac{\partial^2}{\partial x^2} + \beta_2^2 \frac{\partial^2}{\partial y^2}\right) \sigma_x = 0 \tag{2}$$

This Laplace equation also arises by neglecting ε_y, the transverse direct strain. If u and v are the displacement components, $\varepsilon_y = \partial v/\partial u = 0$ shows that $v = v(x)$ only.
Neglecting Poisson ratio effects, the shear strain $\gamma_{xy} = \frac{\partial u}{\partial y} + \frac{dv}{dx}$ leads to

$$\frac{\partial}{\partial x}\left(\tau_{xy} - G_{xy}\frac{dv}{dx}\right) = \beta_2^2 \frac{\partial \sigma_x}{\partial y} \tag{3}$$

where τ_{xy} is the shear stress. One equilibrium condition gives

$$\frac{\partial \sigma_x}{\partial x} = -\frac{\partial \tau_{xy}}{\partial y} = -\frac{\partial}{\partial y}\left(\tau_{xy} - G_{xy}\frac{dv}{dx}\right) \tag{4}$$

and (3) and (4) are Cauchy-Riemann relations in the complex variable $z_2 = \beta_2 x + iy$. Then we may write

$$\beta_2 \sigma_x - i\left(\tau_{xy} - G_{xy}\frac{dv}{dx}\right) = G_{xy}\frac{d}{dz_2}\omega_2(z_2) \tag{5}$$

The general solution of (1) is $\chi = \mathrm{Re}\{F_1(z_1) - F_2(z_2)\}$.

We can use (5) to estimate $F_2(z_2)$, relating local stresses to applied loads. This solution satisfies only some local boundary conditions and $F_1(z_1)$ is then calculated locally to satisfy the remaining boundary conditions. The procedure is illustrated later.

FLEXURE OF BEAMS

All engineering beam theories assume $\varepsilon_y = 0$ and so (5) gives complete solutions. Solutions can often be built up from sums of the elementary solution

$$\sigma_x = \sigma_0 \sin \lambda x \sinh \frac{\lambda y}{\beta_2} \quad , \quad \sigma_x = \sigma_0 \sinh \lambda' x \sin \frac{\lambda' y}{\beta_2} \tag{6}$$

The condition $\tau_{xy} = 0$ on $y = \pm h/2$ leads to expressions for τ_{xy} and v from (5).
The bending moment $M = \int_{-h/2}^{h/2} y\,\sigma_x\,dy$ also follows.

For a rectangular beam of breadth b, integrating (3) between $y = \pm h/2$ gives

$$\frac{dS}{dx} - A\,G_{xy}\frac{d^2 v}{dx^2} = 2b\,\beta_2^2\,\bar{\sigma}_x \tag{7}$$

where S is the sectional shear force $(= \frac{dM}{dx})$, A is the cross-sectional area $(= bh)$, $\bar{\sigma}_x$ is c_x $(\bar{y} = h/2)$.

The usual beam theory assumption $(\sigma_x = ky)$ relates $\bar{\sigma}_x$ to M and (7) becomes

$$\frac{h^2}{12\beta_2^2} \frac{d^2 M}{dx^2} = E_x I \frac{dv^2}{dx^2} + M \qquad (8)$$

where I is the sectional second moment of area $(= bh^3/12)$.

The right hand side of this equation gives the Euler-Bernoulli beam equation and the left hand side is closely related to the Timoshenko shear correction. More generally, for a known non-linear dependence on y — such as in (6) — a shape factor α can be introduced,

$$\alpha = \frac{6}{h^2} \int_{-h/2}^{h/2} y \frac{\sigma_x}{\bar{\sigma}_x} dy \qquad (9)$$

the multiplying constant being chosen to give $\alpha = 1$ for the case when the longitudinal direct stress σ_x is linearly dependent on y, and (8) becomes

$$\frac{h^2}{12\beta_2^2} \frac{d^2 M}{dx^2} = E_x I \frac{d^2 v}{dx^2} + \frac{M}{\alpha} \qquad (10)$$

Two simple cases give α = constant along the beam. It is found that, for a pinned beam of length π/λ subjected to transverse loading $p = p_0 \sin \lambda x$, the central deflection is

$$v_0 = \frac{1}{2\alpha} \coth \frac{\mu}{2} \cdot \frac{p_0}{\lambda^4 E_x I} \quad , \quad \alpha = \frac{6}{\mu} (\coth \frac{\mu}{2} - \frac{2}{\mu})$$

where $\mu = \lambda h/\beta_2$ (proportional to aspect ratio). Figure 1 shows, for a given beam with given Young's modulus, the increase in central deflection with reducing shear modulus.

The free vibration of a pinned beam also gives this form of loading since now $p = \rho A (\partial^2 v/\partial t^2)$, where ρ is the mass/unit length.

With a natural frequency $\omega/2\pi$ and defining a non-dimensional frequency $\zeta = \frac{h\omega}{\beta_2} (\rho/E_x)^{1/2}$, Fig. 2 compares the solution for this problem with the result obtained from Euler-Bernoulli theory, $\zeta = \mu^2/(12)^{1/2}$. The solution is very close to that obtained using the Timoshenko shear correction.

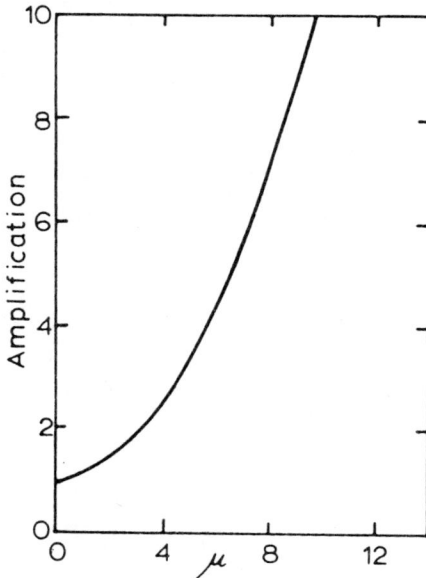

Fig. 1 Effect of the parameter μ on the central deflection of a beam under sinusoidal loading

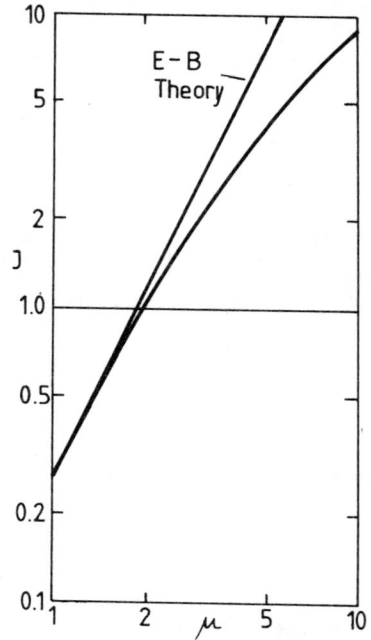

Fig. 2 Comparison of natural frequency of pinned composite beam with Euler-Bernoulli theory.

In general, however, α is a function of x. Some progress can be made by assuming an "equivalent" constant value of α (say α_o) and estimating it a posteriori. For free vibrations of a free-free beam with ends at $x = \pm \ell/2$, we substitute

$$v = V_o \exp(i\omega t) \quad \text{where} \quad V_o = V_\lambda \cos \lambda x + V_{\lambda'} \cosh \lambda' x \quad (11)$$

This example is treated in detail in 2 . We merely sketch the analysis here. From the boundary conditions,

$$\lambda \tan \frac{\lambda \ell}{2} + \lambda' \tanh \frac{\lambda' \ell}{2} = 0 \quad (12)$$

while, from (10), we derive the result

$$\frac{1}{\lambda'^2} = \frac{h^2 \alpha_o}{12\beta_2^2} + \frac{1}{\lambda^2} \quad (13)$$

Substituting (11) into (5), σ_x, τ_{xy} and thence M can be found.
With $M = M_o \exp(i\omega t)$, $M_o = M_\lambda \cos \lambda x + M_{\lambda'} \cosh \lambda' x$, (9) gives

$$\alpha(x) = \frac{M_\lambda \cos \lambda x + M_{\lambda'} \cosh \lambda' x}{M_{\lambda}/\alpha_\lambda \cos \lambda x + M_{\lambda'}/\alpha'_\lambda \cosh \lambda' x} \tag{14}$$

where α_λ and $\alpha_{\lambda'}$ are the values of α for the two elementary solutions (6).
To estimate α_o, we postulate that

$$\frac{1}{\alpha_o} \int_o^{\ell/2} M_o \, dx = \int_o^{\ell/2} M_o/\alpha \, dx$$

(which ensures that the exact and approximate solutions have the same tip
slope). From (14), this gives :

$$\alpha_o = (\frac{1}{\mu^2} + \frac{1}{\mu'^2}) \quad (\frac{1}{\alpha_\lambda \mu^2} + \frac{1}{\alpha'_\lambda \mu'^2}), \quad \mu' = \lambda' \, h/\beta_2 \tag{15}$$

Iteration between (12), (13) and (15) readily yields λ, λ', a_o, etc.
The departure from the Euler-Bernoulli values is much as for the pinned
beam and again justifies the Timoshenko shear correction.
Figure 3 shows the variation of α along the beam for the fundamental vibration, with $\beta_2 = 0.16$ and aspect ratio 6. The numerator in (14) vanishes
at $x = \ell/2$ due to the boundary condition $M = 0$ there, giving $\alpha = 0$. As a result, since $\alpha_{\lambda'} > \alpha_\lambda$, the denominator vanishes at some value of $x < \ell/2$,
corresponding to the condition $\bar{\sigma}_x = 0$ there.
The negative values of α near the tip indicate stress reversal, $\bar{\sigma}_x$ and M taking
opposite signs ; this is necessary to secure $M = 0$ at $x = \ell/2$.

Although the solution outlined above satisfies $M = 0$, $S = 0$ on the end faces,
it does not give vanishing σ_x, τ_{xy} over these faces.
For an isotropic beam, St. Venant's principle can be invoked to justify this
discrepancy but, for an anisotropic beam, the form of (2) shows that the
disturbance is extended along the beam in the ratio β_2^{-1}. In the example quoted
the disturbed area is not limited to the tips of the beam but covers the
entire length. Analysis shows that the correction to the stress distribution
at $x = 0$ is of the order of $1/2$ % and so has little effect on the natural frequency. However, it may well be that the stress reversal predicted will occur
in practice.

This difficulty also arises with other forms of support and loading. For
example, the method of clamping a cantilever determines the stress distribution
at this section. This has an effect on the stress distribution away from
the clamped end which manifests itself through the appearance of the parameter
in (10). Figure 4 shows three configurations the right hand ends of which
represent the same cantilever.

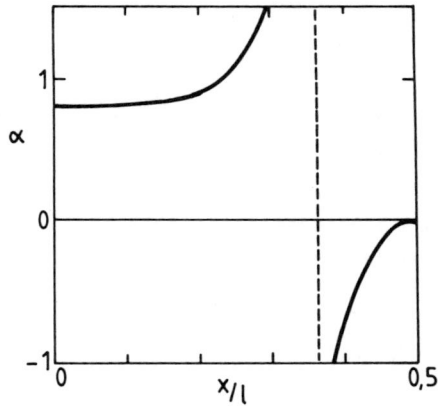

Fig. 3 Variation of α along a beam in free-free vibration.

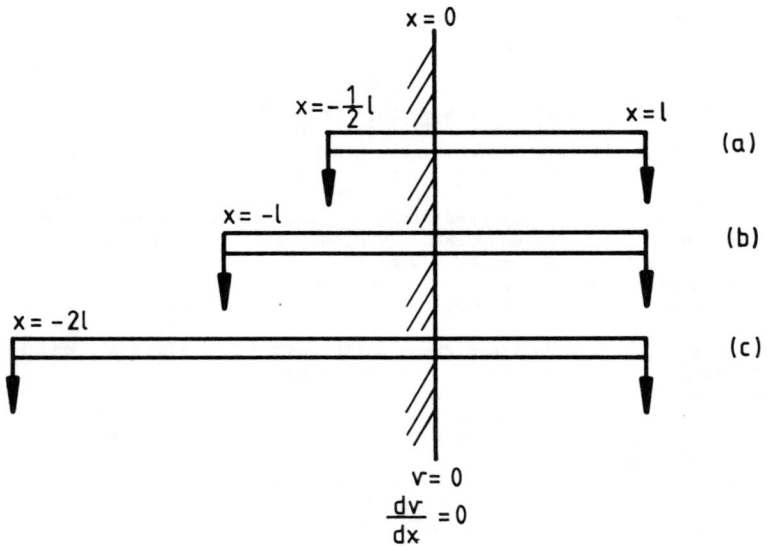

Fig. 4 Beams in three-point loading with the same cantilever at the right hand end.

Since $\alpha(x)$ will be different in each case, the values of tip deflection will differ. In principle, this can be found from (10) since M can be expressed as a known Fourier series in x.

However, convergence is slow since the point load at $x = 0$ gives a discontinuity in dM/dx. This has been tackled by direct numerical solution of (2) ; we find a spread of 10 % in tip deflection over the three cases for an aspect ratio of 10 with $\beta_2 = 0.16$.

PROBLEMS OF CRACKS AND NOTCHES

The exact form of solution at the root of a notch is found from (1). Stresses take the form $A_1 z_1^{-n} + A_2 z_2^{-n}$ to satisfy the conditions of vanishing shear stress and normal direct stress on the notch surfaces. Figure 5 plots n for a symmetrical notch of semi-angle $(\pi/2 - \alpha)$ for $\beta_1 = 1.13$, $\beta_2 = 0.16$. It can be shown that, for the simple solution of (2), the logical boundary condition is the vanishing of the normal derivative of u in the z_2-plane. Figure 5 also shows the value of n obtained for the simple solution, adequate agreement being obtained. The simple problem is readily solved for many geometries thus relating A_2 to the applied loads. We then determine A_1 to ensure vanishing of the shear stress on the surfaces of the notch. All continuum solutions of notch problems lead to stress singularities at the root ; this is one reason why fracture mechanics invokes an energy criterion. However, by assuming constant stresses across the fibre and matrix layers of a composite, estimates can be made of finite maximum stresses at the root of the notch, these values depending on volume fraction and fibre dimensions.

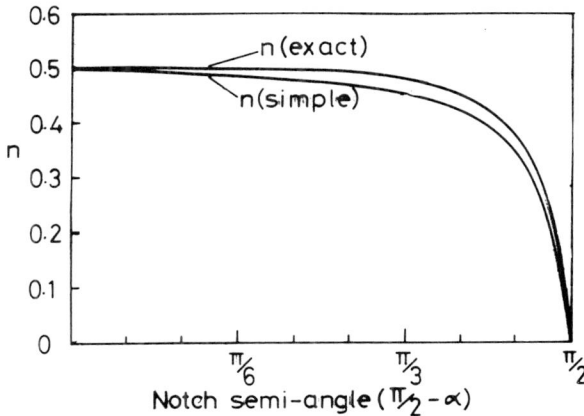

Fig. 5 Dependence of n on notch semi-angle for the case $\beta_1 = 1.13$, $\beta_2 = 0.16$.

We illustrate for the most complicated case — that of the longitudinal direct stress σ_x. If u_o, u_1, u_{-1} represent the longitudinal displacements in the fibres nearest the notch root (see Fig. 6), we replace (2) (which also governs u) by the differential - difference equation

$$\frac{d^2 u_o}{dx^2} = \frac{\beta_2^2}{(h_f + h_m)^2} (2u_o - u_1 - u_{-1})$$

where h_f, h_m are the thicknesses of fibre and matrix layers.

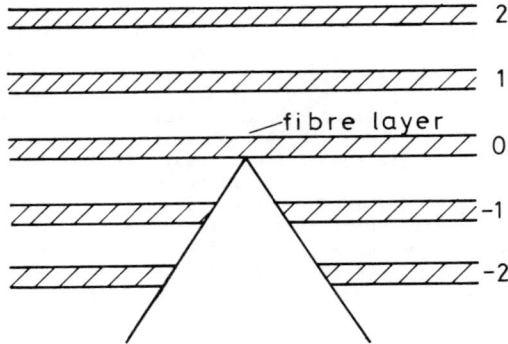

Fig. 6   Notation for fibre layers

We can express this as

$$\frac{d^2 u_o}{dx^2} - p^2 u_o = F(x) \; , \quad p^2 = \frac{2\beta_2^2}{(h_f + h_m)^2}$$

where $F(x)$ is found from the known continuum solutions. Then, to good accuracy,

$$\sigma_{x,o} = \int_{-\infty}^{o} F(x) \, e^{px} \, dx$$

Applying this to a normal crack of depth h, we find

$$\sigma_{x,max} = \sigma_o \left\{ 1 + 1.0531 \left(\frac{h}{h_f + h_m} \right)^{1/2} \right\}$$

Results from this can be compared with experiment as reported by St. John and Street at the 3rd International Congress on Fracture, 1973, for an aluminium-boron composite. Figure 7 shows a reasonable comparison bearing in mind that the experimental material is ductile, yielding near the crack tip. G.R.P. usually fails in the matrix and theoretical estimates are compared with experiment in Fig. 8 for symmetrical notches of various semi-angles.

COMPARISON WITH THEORY FOR IDEAL COMPOSITE

An ideal composite is defined as one which is incompressible and whose
fibres are inextensible. The former property imposes conditions on the
Poisson ratios and, in terms of the notation of this paper, implies for
plane strain problems (to which the theory is virtually exclusively applied
in two-dimensional problems) that $\beta_1 \to \infty$.
Inextensibility implies that $\beta_2 = 0$. This theory leads to line forces
(lines of infinite stress) acting along certain fibres or perpendicular to
the fibre direction. It is modified in the region of these line forces by
a boundary layer approach, developed in terms of the small parameters β_2
and β_1^{-1}, the former being relevant along the fibres and the latter at right
angles to them.

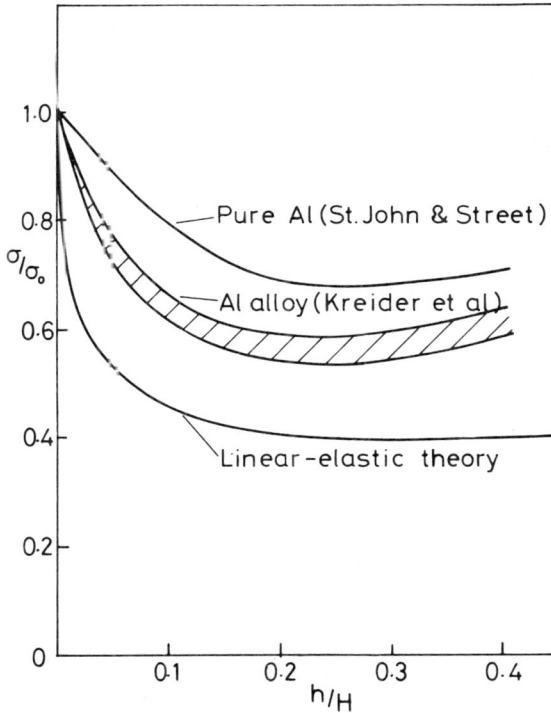

Fig. 7 Net section stress for fracture vs. crack depth for an aluminium-
boron composite.

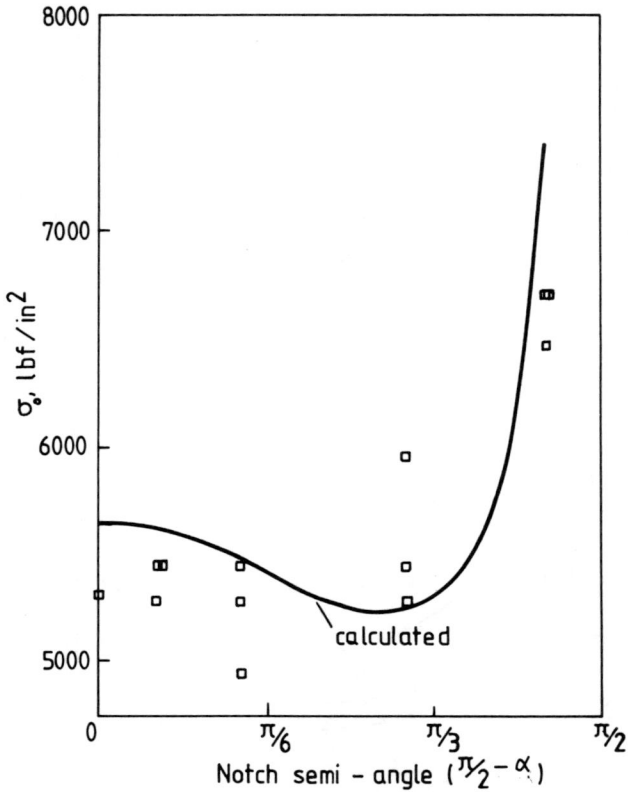

Fig. 8 Tensile load for failure vs. notch semi-angle for glass-mat G.R.P.

The theory of the present paper merely regards β_1 and β_2 as finite material properties, though it does require both β_2 and β_2/β_1 to be small. Its basic premise is that, in particular loadings of particular specimens, the transverse displacement gradient $\partial v/\partial y$ is negligibly small except in small regions identifiable a priori. This is in contradistinction to the ideal theory which states that the postulates of inextensibility and incompressibility apply over most of the specimen.

Both theories require β_1 and β_2 to be estimated from theory or measurement. Either way there are problems. Theoretical estimates using the law of mixtures are known to be only approximate (except possibly for E_x). On the other hand, measurements requires evaluation of five elastic constants for the simplest case of a transverse isotropic composite, and more for other

other geometries, and it is known how to do this. Thus the elastic properties of the material are imperfectly known and this limits the accuracy of any theoretical analysis. The main purposes of theory are therefore to obtain order-of-magnitude estimates, which in many situations is all that is required, and to give physical insights.

Despite this limitation, there is a strong need for evaluating these theories in order to assess when, and for which materials, they might be of value. One way is by a systematic experimental programme ; another is by comparison with exact solution of (1), which can only be determined by computer, except in a handful of cases.

In the meantime, some comparison of the two theories can be made in the two problem areas considered in this paper. The flexure of beams is discussed in [3] for ideal beams. It is seen to apply to cantilevers whose aspect ratios lie in the range : $-\beta_1$ < aspect ratio < $1/2 \ \beta_2^{-1}$. The lower limit is dictated by the requirement that adjustment to the applied shear load on the end face shall be accomplished over a proportionally small length of beam at that end — a recognition of the problem with St. Venant's principle. The upper limit is required in order that the boundary layers associated with the line forces on the top and bottom faces of the beams shall be separate. This latter requirement is not applicable to the theory of this paper since $\nabla_2^2 \ \chi = 0$, the governing boundary layer equation of the ideal theory, is now the governing equation for the entire beam. In fact, the assumption of negligible $\partial v/\partial y$ implies that the present theory only applies for aspect ratios in excess of $1/2 \ \beta_2^{-1}$. For lower aspect ratios, the contribution to strain energy due to transverse stresses is not negligible. We note that, for the case of a uni-directional C.R.F.P. composite for which experiment gives $\beta_1 = 1.13$, $\beta_2 = 0.16$, ideal theory applies for aspect ratios between 1 and 3 while the present theory applies for aspect ratios greater than 3. Thus the present theory appears to be applicable for more practical aspect ratios.

The ideal theory has been applied to some problems of cracks (see [4] and [5]). In these papers inextensibility, i.e. $\beta_2 = 0$, applies over most of the specimen (the outer solution) whereas, in the present theory, a solution $\nabla_2^2 \ \chi = 0$ applies. Both theories use the full solution involving β_1 and β_2 as a local solution at the crack tip, (in fact, for ideal theory, [5], using β_1- and β_2-terms, obtains the same results as [4] using β_2-terms only for the inner solution), ideal theory using it to match the line force arising from the main field solution and the present theory using the β_1-term to modify the β_2-term of the main field solutions. In the case where an exact solution of (1) is known - the case of a crack normal to the fibre direction in an infinite plate (see [1]) - it is found that, at great distances from the crack, the contribution of the ∇_1^2 solution is β_2/β_1 times that of the ∇_2^2 solution, indicating the dominance of the latter term as postulated by the present theory.
For finite plates, the two theories give different results for the dependence of the magnitude of the singularity on the ratio of crack length to specimen length. The ideal composite theory gives results independent of the width of plate whereas the present theory gives the well-known Irwin correction as for isotropic materials.

A further important difference between the two theories is that the present theory is limited to two-dimensional problems whereas the ideal composite theory can be applied to more complicated specimens and lay-ups, including filament winding and cross-ply lay-ups.

Thus although it is believed that the present theory offers more realistic results for a limited range of problems, the ideal composite theory offers the only solution available in a wide range of situations.

REFERENCES

1. King, J.L., Some aspects of continuum mechanics applied to structured materials. Proceedings of the International Symposium on the Mechanical Behaviour of Structured Media, Ottawa, 1981. (Published by Elsevier as Mechanics of Structured Media, Studies in Applied Mechanics, 5A).

2. King, J.L., The free transverse vibrations of anisotropic beams. To be published in Journal of Sound and Vibration, 1985.

3. Everstine, G.B. and Pipkin, A.C., Boundary layers in fiber-reinforced materials. Journal of Applied Mechanics, 40, pp. 518-522, 1973.

4. Sanchez-Moya, V. and Pipkin, A.C. , Crack tip analysis for elastic materials reinforced with strong fibres. Quarterly Journal of Mechanics and Applied Mathematics, 31, pp. 349-362, 1978.

5. Spencer, A.J.M., On the stress-intensity factor for a crack in a highly anisotropic plate. Quarterly Journal of Mechanics and Applied Mathemathics, 32, pp. 437-444, 1979.

LIFE-TIME PREDICTION OF THICK COMPOSITE SHELL STRUCTURES

A. BOUFERA, P. HAMELIN

Centre d'Etudes et de Recherches sur les Materiaux Composites (CERMAC)
Laboratoire Bétons et Structures, INSA, Villeurbanne, France.

INTRODUCTION

Composite materials are particularly valorised in industrial applications since they prove multifunctional properties. The gain of weight for such materials is more significant in case of immersed sub-marine structures at great depths (6000 m), which request to take into consideration Archimedes pressure.

Corrosion, stability, electrical and magnetic properties are favourable factors for composite formulations. The first table gives us an elementary comparison between different materials.

Table 1

Materials	Weight in air	Weight in water
Steel	7,8 kg	6,8 kg
Titane	4,5 kg	3,5 kg
Composite	2,0 kg	1,0 kg

The purpose of our study is to define a container described by Fig. 2.

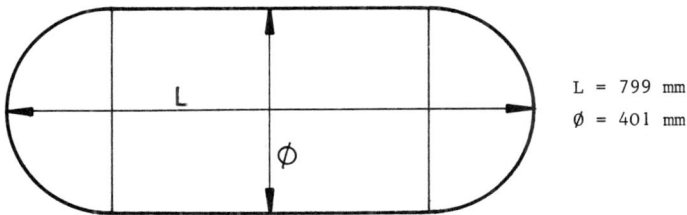

L = 799 mm
$\emptyset$ = 401 mm

Fig. 2 Description of a container.

The cylindrical part of the container is made by filament winding (glass E-epoxy resin) and the hemispherical parts are moulded by a low pressure press at low temperature. The external anticorrosion layer is made of vinyl-ester resin. The different elements are joined together by degraded multi-layers.

The length of the container is 799 mm, the internal diameter is 401 mm, the radius of the sphere is 200 mm. The volume of the structure is 100 ℓ.

Using a compressive strength of 450 MPa and a Young modulus of 23000 MPa for a simplified calculation (1,2) described in Table 3, we can project that the thickness of the cylindrical part could be 52 mm and the thickness of the sperical part could be 33 mm.

We can consider that the ratio between thickness of the cylinder and its diameter is 0,12. The formulation of the composite material of the cylinder is desequilibrated : 458,2 gr/m² of glass in the chain direction and 279,6 g/m² of glass in the woof direction for one layer. The formulation of the composite for the sphere is equilibrated : 845 g/m². Matrix are epoxies resins with an anhydrid hardener for the cylindrical part and a cyclo-aliphatic for the spherical part.

Table 3 Elementary calculations.

CYLINDER	EXTERNAL LOAD = INTERNAL STRENGTH	$e = 52$ mm 80 LAYERS
	$q_{crit} = \dfrac{E\,l}{R}\cdot\dfrac{1}{n^2\frac{1}{2}(n\frac{l}{L})^2}\left\{\dfrac{1}{[n^2(\frac{l}{\pi m})^2 n]^2}+\dfrac{h^2}{12\,R^2(1-\nu^2)}[n^2\frac{l}{L}]^2\right\}$	$P_{CR} = 980\ b$
SPHERE	EXTERNAL LOAD = INTERNAL STRENGTH	$e = 33$ mm 47 LAYERS
	$q_{crit} = \dfrac{2\,E\,e^2}{R^2\sqrt{3(1-\nu^2)}}$	$P = 5300\ b$

A preceeding study (3) of J.M. Reynouard and P. Hamelin has taken into consideration the influence of the ratio thickness/ radius, and the influence of the shear stress in such thick shell.

Different methods of calculations have been tested (isoparametric element brick with twenty nodes, degenerate isoparametric element with two nodes in the thickness,...) and the principal factors affecting the mechanical behaviour of such structures were the difference of normal stress between the internal and the external faces and the difference of stress for the cylinder between the longitudinal direction and the circumference.
In any case, the buckling effect is preponderant but it is the compressive behaviour of the multi-layers which is the main parameter.

In conclusion, we can consider that the life-time of such shell is function of the rheological characteristics of the materials and classic numerical methods are adequate for an evaluation of the stresses, strains and displacements at each point of the structure.

RHEOLOGY AND BEHAVIOUR LAWS UNDER COMPRESSIVE LOAD

Strength-strain relation
Experimental curves (Fig. 4) are drawn from cubic specimens and rectangular specimens loaded with a displacement regulation, or a load regulation.
If we consider the fracture mechanisms, we can observe that the failure mode is characteristic :

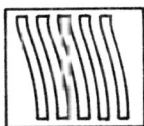

shear and buckling
mode

layer dislocation under
shear effect

In function of the compressive strength we measure the deformation in the direction of the load and the global displacement.

On the Fig. 5, we can see that the deformations on both sides are not symmetrical and we can explain these phenomena by different reasons. First point, the shape of the specimen : We have choosen a cubic model for an evaluation of the ultimate load of the materials in the same conditions of thickness. Second point, the buckling phenomenon induces a flexural effect. Third point, the boundary conditions and the orthotropic characteristics of the materials influence the modes of deformation.

Figure 6 proves these effets. The equilibrated multilayers (V) give the same results with different boundary conditions under compressive loading. On the contrary, the formulation M (60 % of glass in the direction of load, 40% of glass in the perpendicular direction), present two different strain-stress relations.

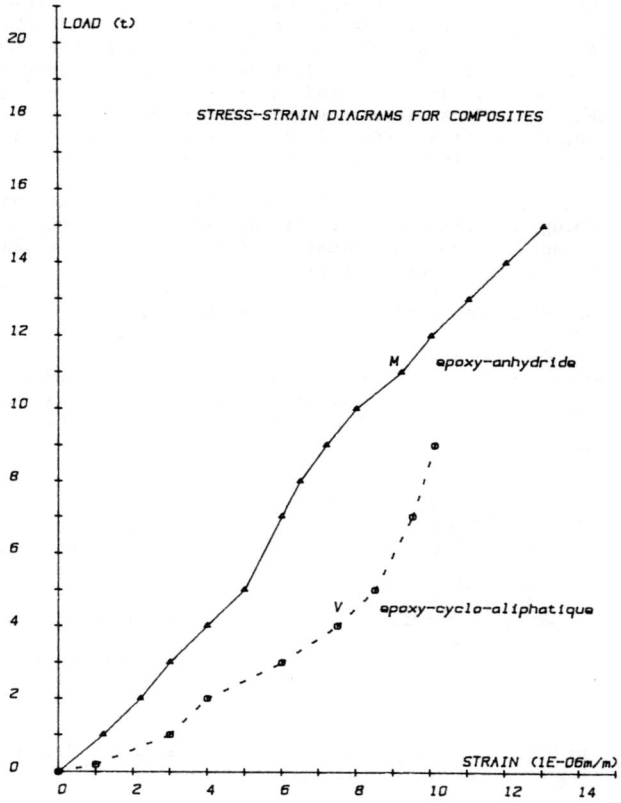

Fig. 4   Stress-strain diagrams for composites

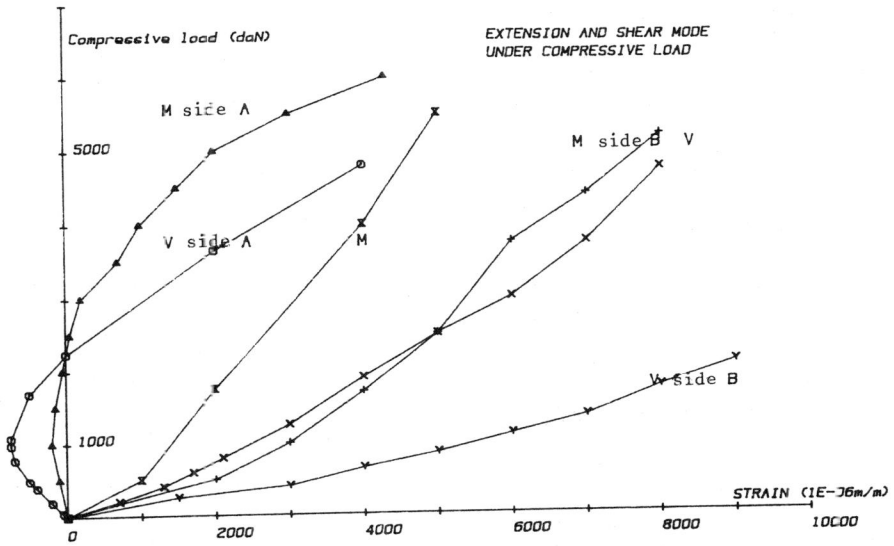

Fig. 5 Extension and shear mode under compressive load.

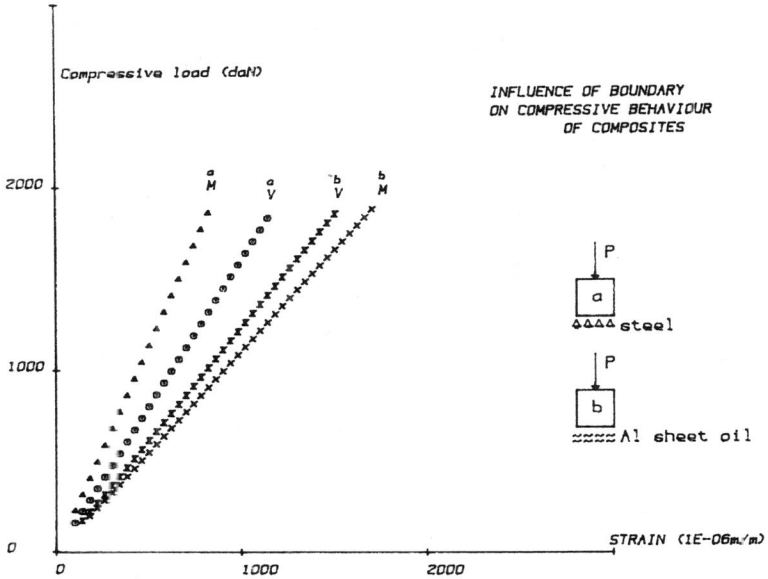

Fig. 6 Influence of boundary on compressive behaviour of composites.

Comparison law's modelisation

Many authors as Dow, Fried, Rosen (4) and Scheurch (5) have studied compressive failure. They considered failure of a composite consisting of a set
of parallel plates under edge loading, analyzed two limiting modes of buckling
failure by energy methods and obtained similar results.

Consider parallel fibers (Fig. 7), represented as plates of thickness h
and length L, each subjected to compressive load P and separated by a matrix
of dimension 2e. Two buckling modes may be visualized : adjacent fibers
buckle in opposite direction, or in phase. The first mode is called the
extension mode, the second a shear mode.

EXTENSION **MODE**	$\sigma_{f_{crit}} = 2\left[\dfrac{R.E_m.E_f}{3(1-R)}\right]^{\frac{1}{2}}$ $\sigma_c = R\,\sigma_{f_{crit}}$ $\varepsilon_{crit} = \dfrac{\sigma_{f_{crit}}}{E_f}$	$\sigma_c = 2k_f\sqrt{\dfrac{k_f.E_m.E_f}{3(1-k_f)}}$ $\sigma_c = \dfrac{G_m}{1-k_f}$	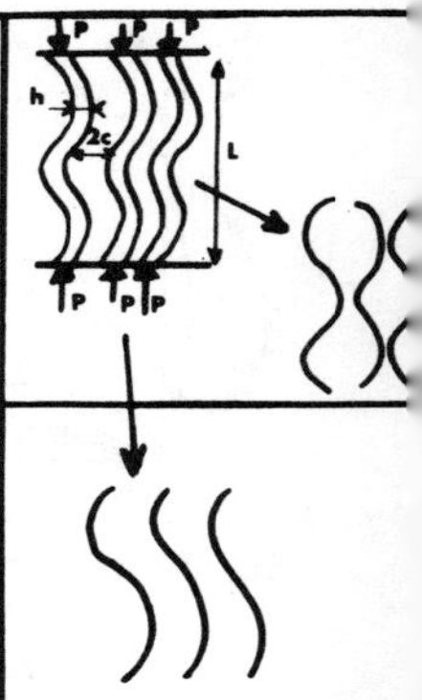
SHEAR **MODE**	$\sigma_{f\,crit} = \dfrac{G_m}{R(1-R)}$ $\sigma_c = \dfrac{G_m}{1-R}$ $\varepsilon_{crit} = \dfrac{1}{R(1-R)}\cdot\dfrac{G_m}{E_f}$	$\sigma_c = a\,\tau_{12} + b$ $\sigma_c = 23{,}8 + 13{,}4\,\tau_{12}$ $^{(hb)}$ $^{(glass)}$ $\sigma_c = 29{,}03 + 9{,}04\,\tau_{12}$ $^{(hb)}$ $^{(carbon)}$	
MIXTURE **RULES**	$\sigma_c = \sigma_{mc}\left[\beta_{mc}.k_m + \beta_{fc}.k_f.\dfrac{E_f}{E_m}\right]$ $\sigma_c = Min\begin{cases}\sigma_c = E_f.\dfrac{\sigma_f}{E_f}.C_{mc}\\[2mm]\sigma_c = E_f.\varepsilon_{mc}\end{cases}$		

Fig. 7 Compressive failure.

The expression of calculations of the critical stress in the first case
is function of the equation

$$\sigma_{f_{crit}} = \frac{\pi^2 E_f h^2}{12L^2}\left[m^2 + \frac{24L^4}{\pi^2 ch^3}\frac{E_m}{E_f}\left(\frac{1}{m^2}\right)\right]$$

by the condition

$$\frac{\partial \sigma_{f_{crit}}}{\partial (m^2)} = 0$$

$$\sigma_{f_{crit}} = 2 \left[\frac{R.E_m.E_f}{3(1-R)} \right]^{1/2}$$

$$\sigma_c = R \, \sigma_{f_{crit}}$$

$$R = \frac{h}{h+2e}$$

$$\varepsilon_{crit} = 2 \left[\frac{R}{3(1-R)} \right]^{1/2} \cdot \left(\frac{E_m}{E_f} \right)^{1/2}$$

For the shear instability, we obtain these expressions :

$$\sigma_{f_{crit}} = \frac{G_m}{1-R}$$

$$\sigma_c = \frac{G_m}{1-R}$$

$$\sigma_{crit} = \frac{1}{R(1-R)} \frac{G_m}{E_f}$$

Fried, from an other "energetic-interpretation" gives two expressions of the compressive stress in the case of buckling in function of the volumic percentage of fiber and of the characteristics of each materials

$$\sigma_c = 2K_f \sqrt{\frac{K_f.E_m.E_f}{3(1-K_f)}}$$

K_f = volumic percentage of fibers

E_f = Young's modulus of fibers

$$\sigma_c = \frac{0,63 \, G_m}{1-K_f}$$

E_m = Young's modulus of matrix

G_m = shear modulus

When the mode of failure is due to the degradation of the interface, then followed by micro-buckling of the fiber due to the important difference between the Poisson's coefficients of fibers and resins, it is possible to establish a relation between the compressive stress of the multilayers and the shear stress.

If we consider the regression curve (Fig. 8), we can propose this expression :

$$\sigma_c = 23.8 + 13.3 \, \tau_{12} \quad (h.b)$$

(for epoxy glass composite).

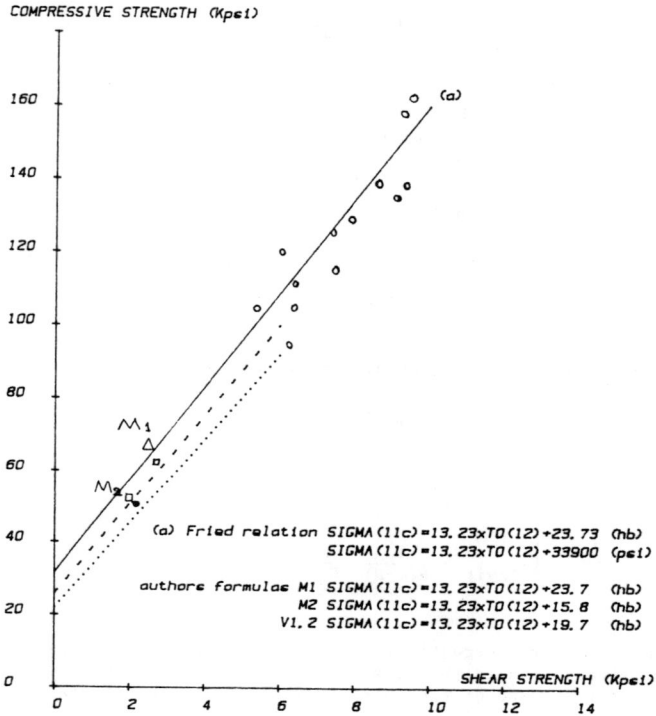

COMPRESSIVE STRENGTH (Kpsi)

(a) Fried relation SIGMA(11c)=13.23xTO(12)+23.73 (hb)
SIGMA(11c)=13.23xTO(12)+33900 (psi)

authors formulas M1 SIGMA(11c)=13.23xTO(12)+23.7 (hb)
M2 SIGMA(11c)=13.23xTO(12)+15.8 (hb)
V1.2 SIGMA(11c)=13.23xTO(12)+19.7 (hb)

SHEAR STRENGTH (Kpsi)

Fig. 8   Correlation between   strength and shear strength of
a composite glass-epoxy.

The last expression is the law of mixtures where the compressive strength can be calculated by :

Chamis expressions
$$
\begin{cases}
\text{(a)} & \sigma_c = E_\ell \cdot E_m \\
\text{(b)} & \sigma_c = E_\ell \dfrac{\sigma_{f_c}}{E_f} C \\
\text{(c)} & \sigma_c = \sigma_{me} \left[\beta_m K_m + \beta_{f_c} \cdot K_f \cdot \dfrac{E_f}{E_m} \right]
\end{cases}
$$

E : longitudinal Young's modulus of the composite ; C : empiric factor ; $\beta_m = 1$ $\beta_f = 0,4$ for glass-epoxy composites.

For the different formulations of composites (M,V) that we have choosen, we can consider (Fig. 8) that the Fried relation, taking into account the mechanical properties of the interface, gives good results for a value of a shear stress τ_{12} of (1,8 - 1,7 MPa) and for a corrective factor function of the percentage of glass.

The other expressions give an evaluation of the ultimate strength different to the experimental values (Table 9).

Table 9

materials	[ROSEN] extension mode	[ROSEN] shear mode	[FRIED] buckling	[FRIED] micro- buckling	mixture laws	experiment values
M_1	4×10^3 MPa	2×10^3 MPa	379×10^3 MPa	142 MPa	a) 49 MPa b) 25 MPa	45 MPa 48 MPa
V	4×10^3 MPa	2×10^3 MPa	261×10^3 MPa	142 MPa	a) 37,50 b) 24 Mpa	35 MPa 42 MPa

The validity of the theorico-experimental expressions is limited to unidirectional composites. If we analyse the modes of failure of our specimens (Fig. 10) made of bidirectional layers (equilibrated and non equilibrated), the buckling effect is limited by fibers present in the perpendicular direction to the load. The method of calculation that we propose, uses formulas for elastic stability of bars (Table 12) in the case : one end fixed, the other end hinged and horizontally constrained over a fixed end. The critical length is the distance between two perpendicular reinforcements.

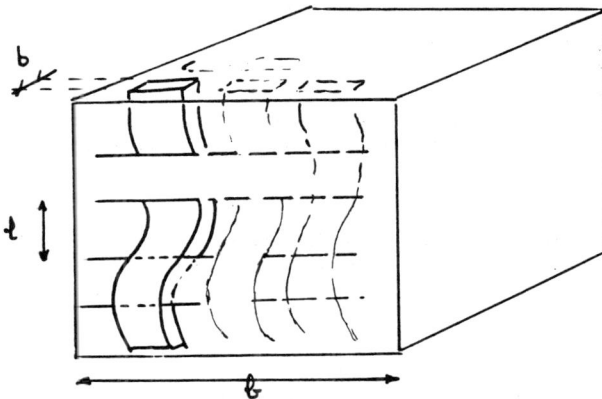

Fig. 10 Failure of composites - Cubic specimen tested.

P_{crit} for one layer :

$$P_{crit} = \frac{\pi^2 b.e^3/12\ E_\ell}{0.7\ \ell^2}$$

b = width of specimen

ℓ = heigth of petch

e = thickness of one layer

$$P_{crit} = 1.17\ E_\ell\ \frac{be^3}{\ell^2}$$

The thickness of the bar is that of one layer and the width that of the specimen tested under compressive load.

Table 12 proves that the limited bucking method is the best formulation for prediction of the ultimate strength in function of the different parameters of bi-directional composites.

Table 12

material	number of layer	thickness of layer (cm)	section (cm)	load of failure (T)	critical load for one layer (daN)	buckling load calculated for one layer (daN)
M_1	31	0,06	2 x 2	19,5	590	$P = 181 \times 10^{-5}\ E_\ell = 600$
M_2	32	0,06	2 x 2	15,2	490	$P = 125 \times 10^{-5}\ E_\ell = 450$
V_1, V_2	30	0,06	1,79 x 1,79	13,6	440	$P = 160 \times 10^{-5}\ E_\ell = 432$

So we can conclude that the compressive strength is function of the Young modulus of the material and function of the different sizes, characteristics and percentage of reinforcement materials. An elementary verification shows that the ratio

$$\frac{\sigma_{c_{chain}}}{\sigma_{c_{tr}}} = \frac{\%\ glass\ in\ the\ direction}{\%\ in\ the\ perpendicular\ direction}$$

RHEOLOGICAL BEHAVIOUR

We study the creeping law or the fatigue law of composites by a rheological analysis of the elastic constants in function of the time and temperature. We apply two methods : the first one consists in analysing thermo-stimulated creeping of each specimen under compressive load ; the second is an application of the viscoelastic theory (9).

Thermo-stimulated creeping

Figure 13 describes the different steps of the experimentation using the
theory of William Landel and Ferry (10).

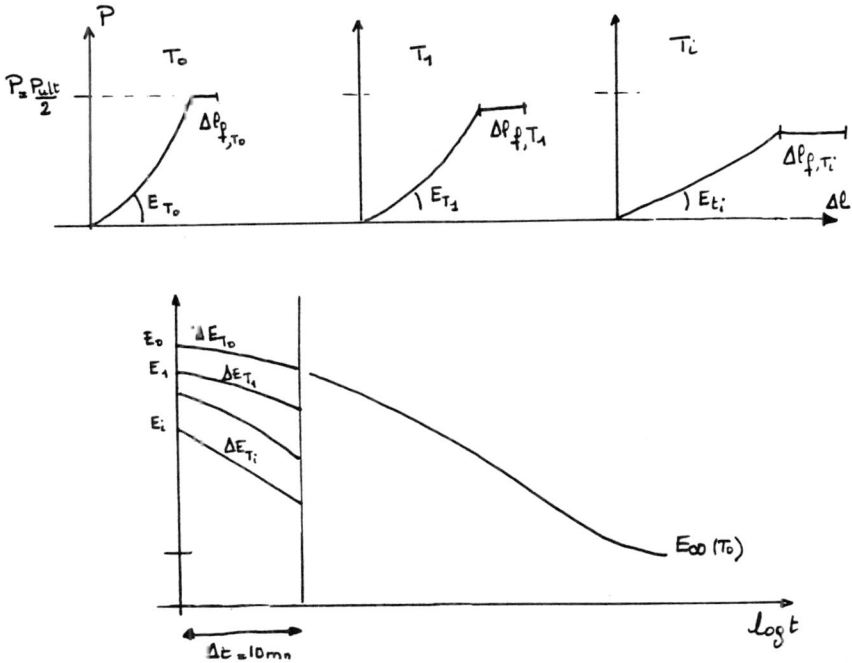

Fig. 13

On the time-temperature equivalence, we can draw a master curve of creeping
effects for a temperature T_0 by translation of each part of curves representing the elastic constant variation of each composite during ten minutes at
different levels of temperature.

For different formulations, we obtain a correction factor (Fig. 14) for
compressive Young modulus either in function of the time, or in function of
the temperature.

For each formulation, we obtain a master curve of creep effect for a temperature of 20°C (Figs. 15, 16).

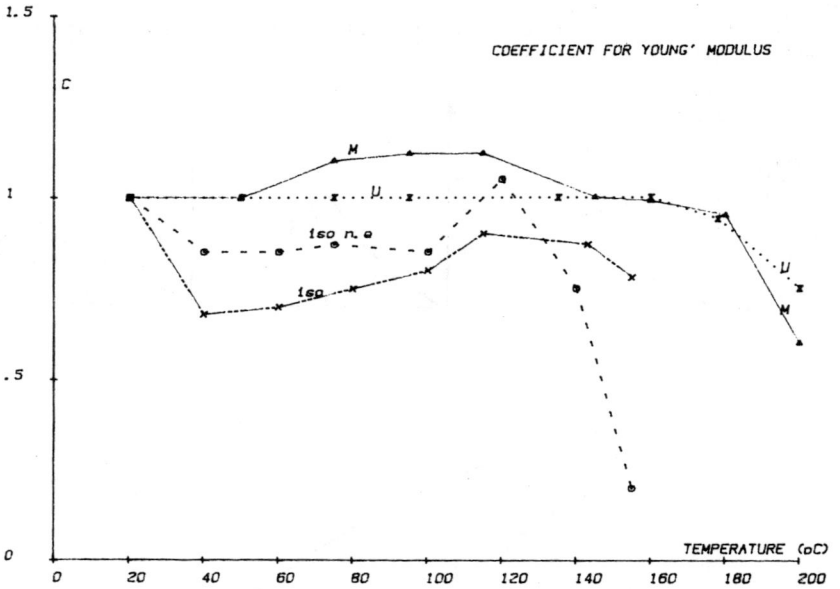

Fig. 14   Coefficient for Young's modulus.

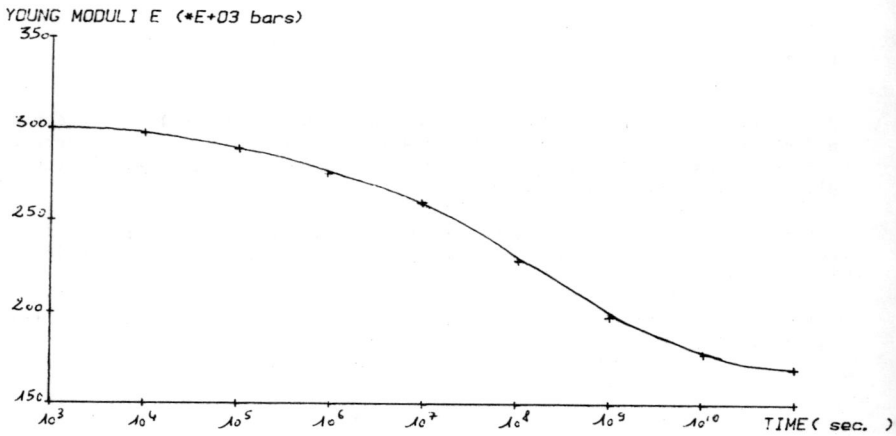

Fig. 15   Relation between Young moduli and time.

RELATON BETWEEN YOUNG MODULI AND TIME

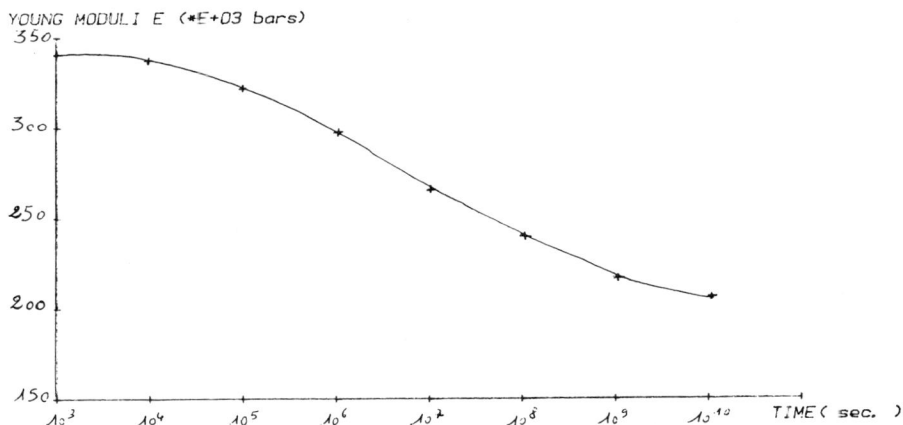

YOUNG MODULI E (*E+03 bars)

Fig. 16 Relaticn between Young moduli and time.

Viscoelastic characteristics and rheological models

The second method consists in using a viscoelastic analyser. In our case we use a Metravib analyser loading the specimen in traction-compression (Fig. 17), or flexion.

From the experimental curves E^*, E', E'', tg δ in functin of the frequency, we can apply the theory of William Landel and Ferry and obtain the curve (20) of the shift factor $a_t = \varphi(H, T_o)$.

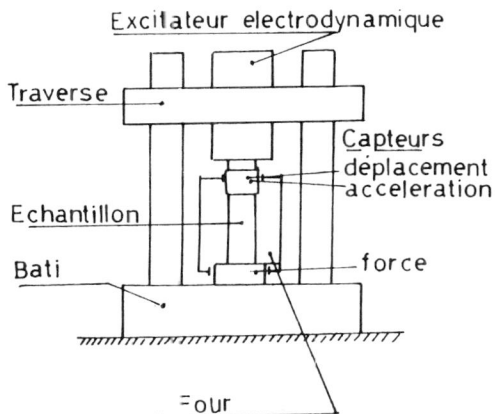

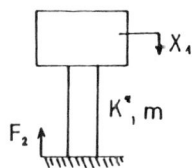

$$\frac{F_4}{X_4} = K^* - m \frac{\omega^4}{6}$$

Fig. 17 Metravib - photoelasticitymeter.

The most interesting figure is the fig. 25 which represents the complex Young modulus $E^* = E_1 + iE_2$ in the complex plane E_1, E_2.

We can observe that all the experimental values are on a same global curve independent of the frequency and of the temperature.

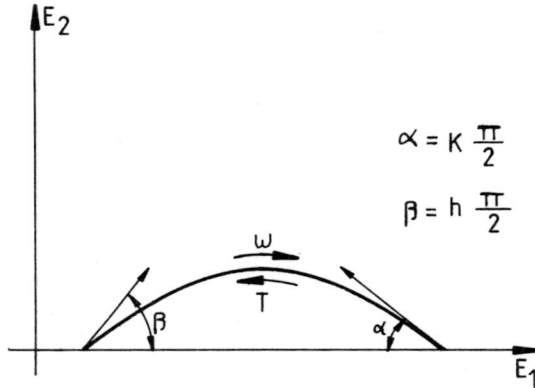

$$\alpha = K \frac{\pi}{2}$$

$$\beta = h \frac{\pi}{2}$$

The rheological model is a parabolic creep model with two parameters : K.S. Cole and R.M. Cole Model (11).
The expression of the complex modulus becomes :

$$E^*(iw) = \frac{E_\infty}{1 + \delta(iw\tau)^{-k} + (iw\tau)^{-h}}$$

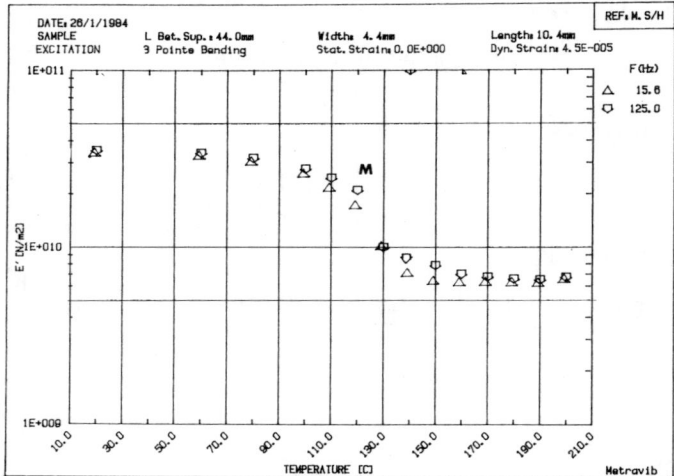

Fig. 18

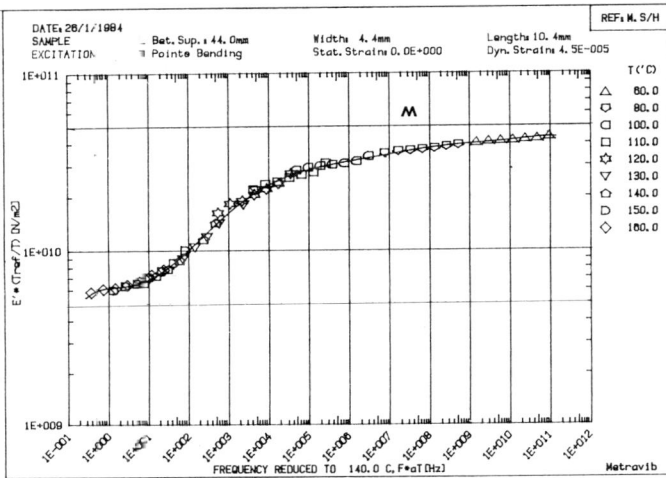

Fig. 19

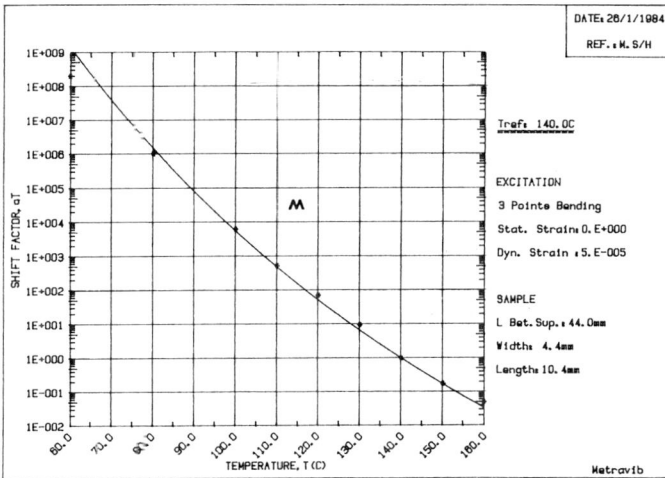

Fig. 20

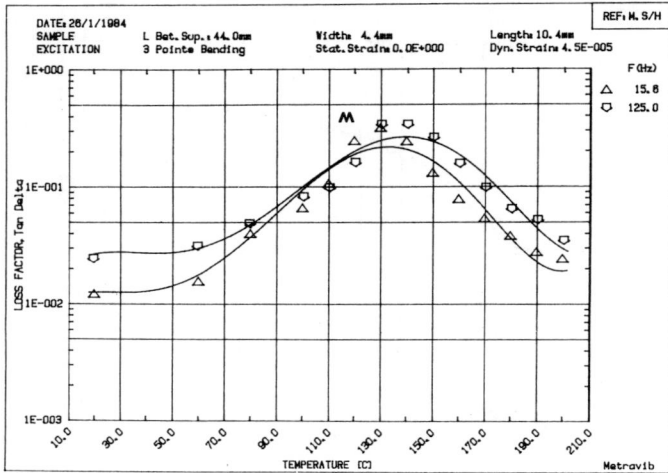

Fig. 21

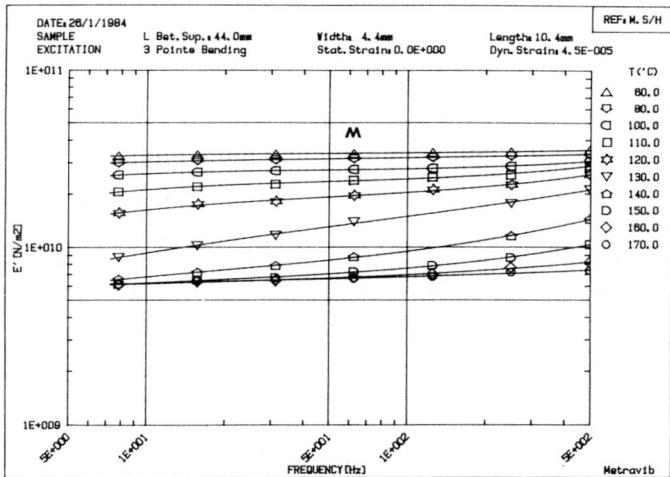

Fig. 22

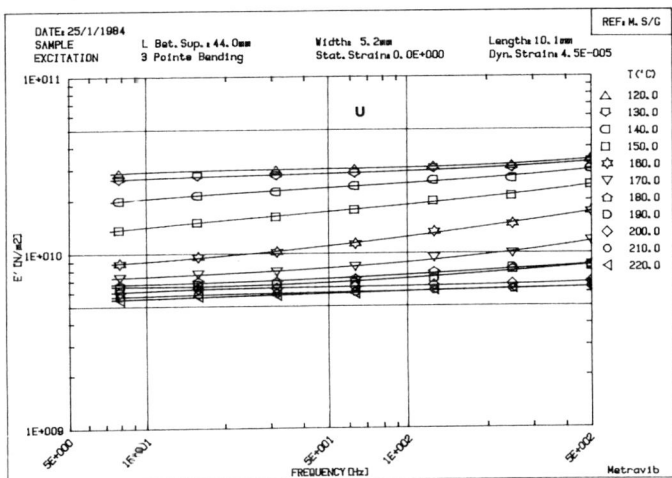

Fig. 23

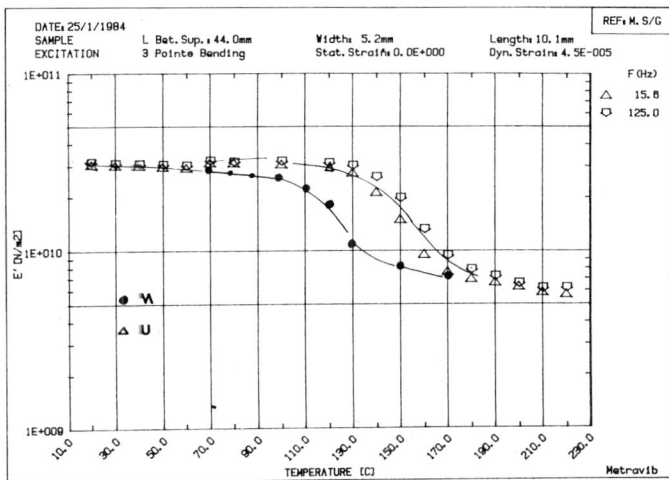

Fig. 24

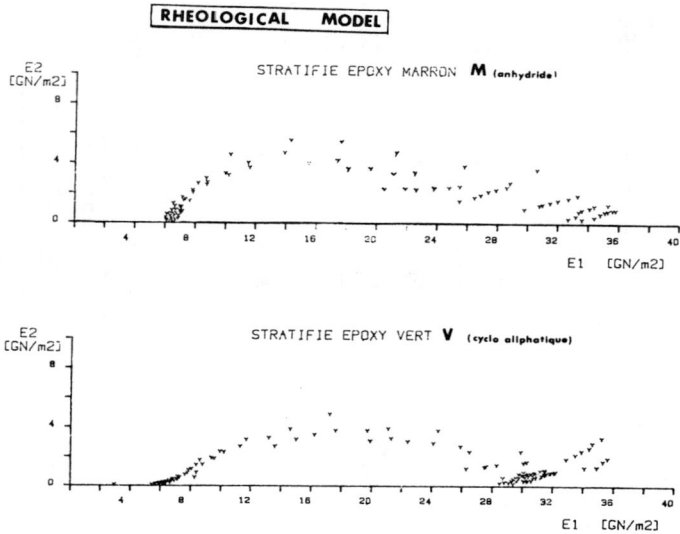

Fig. 25

Applications and conclusions

From the different experimental results, we can establish the table of variation of the mechanical compressive properties (Table 26).

Table 26

Materials	$E_{ostatic}$	$E_{odynamic}$	$E_{5000h(T.C)}$	$E_{5000h(V)}$	$E_{\infty(T.C)}$	$E_{\infty(v)}$
Composite M_1 epoxy-anhybride	35000 MPa	34000 MPa	27000 MPa	27000 MPa	6100 MPa	5000 MPa
Composite V epoxy -cyclo aliphatique	27000 MPa	30000 MPa	25000 MPa	25000 MPa	7400 MPa	6500 MPa

We can deduct what could be the compressive strength in application of the expression of calculation proposed at 1.1 (Table 27).

Table 27

Materials	σ_o MPa	P_{layer}(daN) 5000h	P_{layer}(daN) ∞	$\sigma_{c_{5000h}}$	σ_{c_∞}
composite M_1 epoxy-anhydride	490 MPa	488 daN	110 daN	378 MPa	85,0 MPa
composite V epoxy cyclo-aliphatique	420 MPa	400 daN	120 daN	390 MPa	117 MPa

STRUCTURAL BEHAVIOUR

Evaluations of the strains and the displacements of the container

Using finite element methods and a modelisation of shell elements, we obtain the strain distribution and the radial displacements in each point of the container. For example, Figs. 28, 29 represent the variation of the normal stress along the containter.

The discontinuity between the sphere and the cylinder is visualized and creates a singular point. The variation of stress in each particular section can be drawn (Fig. 30) in function of the pressure. We find a linear law.

Figure 31 represents the distribution of the normal stress for different values of Young's modulus. Figure 32 represents the variation of section of the cylinder for an external pressure of 600 bars.

NORMAL STRESS
CONTRAINTE NORMALE (x. 5) (+) (hbars)

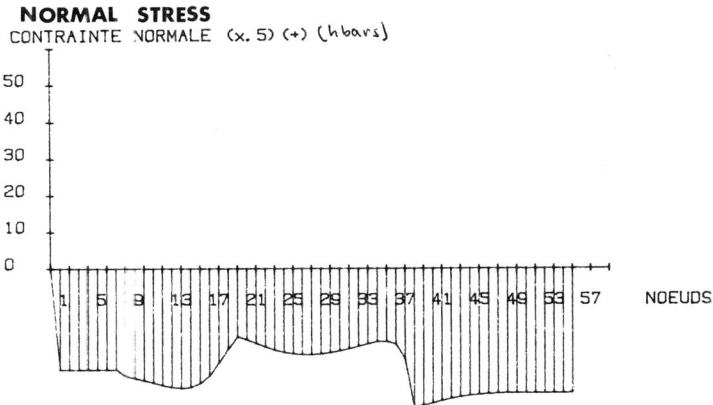

Fig. 28

NORMAL STRESS AT DIFFERENT POINTS OF THE CONTAINER

CONTRAINTE NORMALE (x. 1) (−) (h bars)

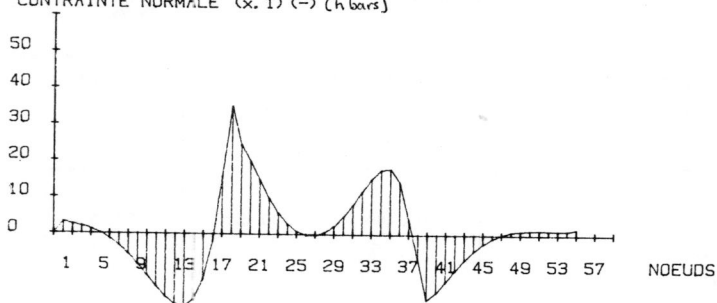

Fig. 29

YOUNG'S MODULI : 1/2 sphere =1330 hbars /Collar = 1700 hbars

POISSON'S COEFFICIENT NU= 0.34

Variation N =F(P)

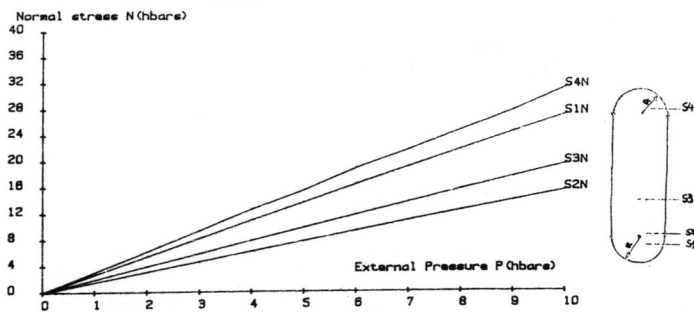

Fig. 30

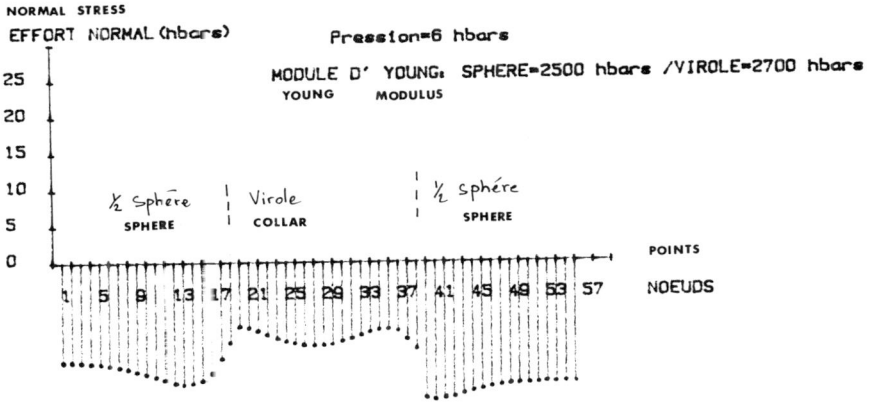

NORMAL STRESS
EFFORT NORMAL (hbars) Pression=6 hbars

MODULE D' YOUNG: SPHERE=2500 hbars /VIROLE=2700 hbars
YOUNG   MODULUS

½ Sphère   Virole   ½ sphère
SPHERE   COLLAR   SPHERE

POINTS
NOEUDS

Fig. 31

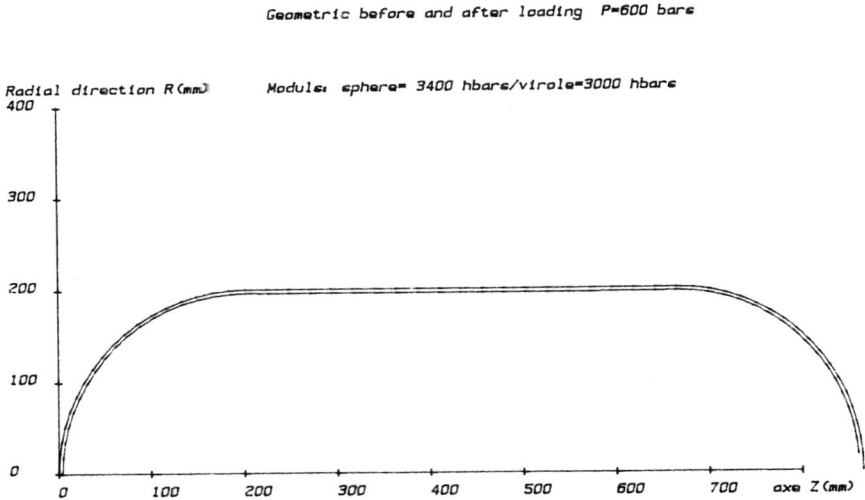

Geometric before and after loading P=600 bars

Radial direction R (mm)   Module: sphere= 3400 hbars/virole=3000 hbars

axe Z (mm)

Fig. 32

Validity of the results of calculations

We have compared the experimental values obtained from an experimentation on a prototype container tested in the high pressure box (1000 bars) of the CNEXO (Brest-France).

Different strain gages have been sticked along the structure on the internal and external faces.
We measured the volume variation in function of the pressure using the principle of communicating vessels (Fig. 34).
The comparison of the experimental values with the results of calculations is satisfying and we find a good concordance between the variation of volume calculated and the measurement of the volume water evicted (Fig. 33).
The tables 35 and 36 examine for different sections and for different values of Young modulus, the stress and displacements in some characteristic sections of the container.

We can consider that the variation of the Young modulus of each material does not change the evaluation of the stress or strain distribution. The principal effect of this variation changes the volume contraction.

CONCLUSION

The life-time prediction for shells made of composite materials loaded by a uniform external pressure is function of the ultimate compressive strength of the composite. For a formulation using a bidirectional glass reinforcement and epoxy resins we have shown that the mode of failure was a buckling mode associated to an over-load of the matrix under a shear stress.

As the ultimate compressive strength of such materials is function of compressive modulus, using a thermo-rheological method, we can anticipate the mechanical behaviour of the structure. For a time of 5000 hours, the security factor is 2, for an infinite time the failure of the container is predicted if we do not change the thickness of each part. The security factor that we have to introduce in an instantaneous calculation is near to 4.

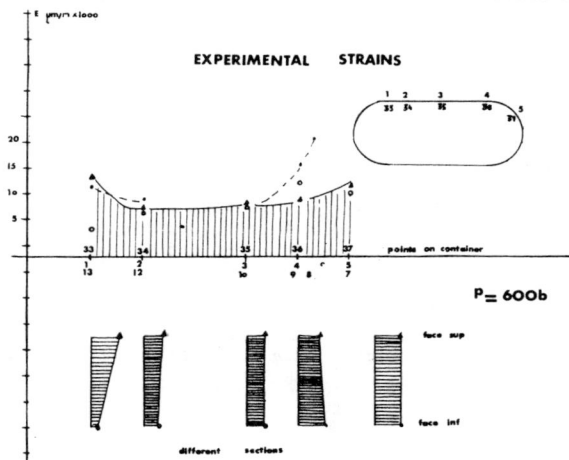

Fig. 33

VARIATION DU VOLUME EN FONCTION DE LA CHARGE

CHARGE [x100 bars]
External pressure

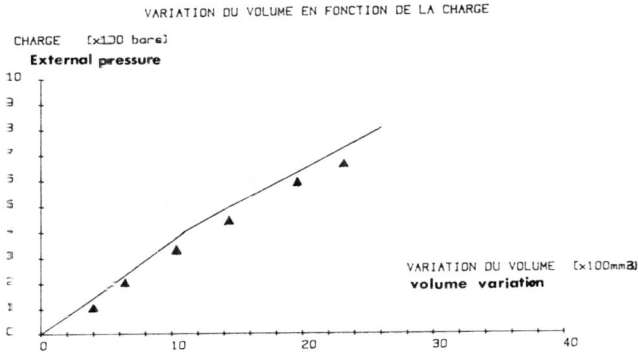

VARIATION DU VOLUME [x100mm3]
volume variation

Fig. 34

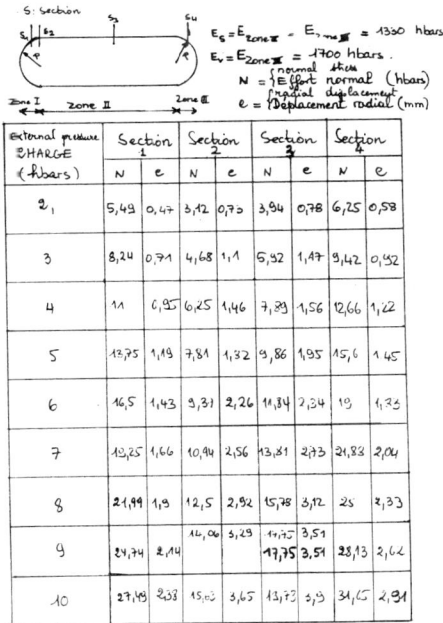

$E_S = E_{zone\,I} = E_{3\,zone\,III} = 1330$ hbars
$E_v = E_{2\,zone\,II} = 1700$ hbars .
$N = \{$ normal stress / Effort normal (hbars)
$e = \{$ radial displacement / Deplacement radial (mm)

External pressure CHARGE (hbars)	Section 1		Section 2		Section 3		Section 4	
	N	e	N	e	N	e	N	e
2,1	5,49	0,47	3,12	0,75	3,94	0,78	6,25	0,58
3	8,24	0,71	4,68	1,1	5,92	1,17	9,42	0,92
4	11	0,95	6,25	1,46	7,89	1,56	12,66	1,22
5	13,75	1,19	7,81	1,32	9,86	1,95	15,6	1 45
6	16,5	1,43	9,31	2,26	11,84	2,34	19	1,33
7	19,25	1,66	10,94	2,56	13,81	2,73	21,83	2,04
8	21,99	1,9	12,5	2,92	15,78	3,12	25	2,33
9	24,74	2,14	16,06	3,29	17,75 3,51 / 17,75 3,51		28,13	2,62
10	27,49	2,33	15,63	3,65	13,73	3,9	31,25	2,91

Fig. 35

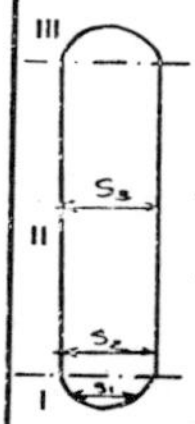

| | YOUNG MODULUS (MPa) | | | P bars | normal stress: N (MPa); displacement max: e (mm) | | |
	I	II	III		S_1	S_2	S_3
III	2500	3500	2500	600	16,3 / 0,7	96,4 / 1,1	119,3 / 1,13
II	2500	2700	2500	600	17,5 / 0,75	83,8 / 1,28	113,4 / 1,48
I	1400	2200	1400	600	159,1 / 1,14	109,5 / 1,84	121,1 / 18
	1400	2200	1400	300	79 / 0,42	50,2 / 0,93	60,2 / 0,9

Fig. 36

REFERENCES

1. Timoshenko, P.S., Théorie de la stabilité élastique, p. 500n n°11-85.

2. Roark, R.Q., Formulas for strain and stress. McGraw-Hill, fourth ed.

3. Hamelin, P., Reynouard, J.M., Containeurs étanches immergeables en matériaux composites. Compte rendu de recherche. Ministère Recherche et de la Technologie, Matériaux GP VI. n°80P.0844

4. Rosen, B.W., Fiber composite materials. Metals Park Ohio, American Society for Metals, 1965.

5. Scheurch, H., NASA report CR 202, Washington DC. Nat. Aeron. and Space Administration, 1965.

6. Chamis, E.C., Micro-structural mechanics and structural synthesis of multilayered composite panels. Report n°9, School of Engineering, Cas Western Reserve University, 1970.

7. Hamelin, P., Contribution à l'étude du comportement rhéologique de liants viscoélastiques en vue de l'analyse du fluage des matériaux composite utilisés sans forme de structures en génie civil. Thèse d'état, Lyon, 1979.

8. Landel W., and Ferry. Viscoelastic properties of polymers. J. Wiley and Sons, New York, 1972, (2nd edition).

9. Cole, K.S. and Cole R.H., Dispersions and absorption in dielectrics. J. Chem. Phys. 0,341, 1941.

THE STABILITY OF COMPOSITE PANELS WITH HOLES

I.H. MARSHALL, W. LITTLE and M.M. EL TAYEBY

Paisley College of Technology, Paisley, Scotland

ABSTRACT

Herein is contained a theoretical and experimental investigation into the effects of circular holes, or cut-outs, on the stability characteristics of thin, rectangular, composite plates. Simply supported square plates with orthotropic material characteristics subject to uniaxial compression are theoretically considered. Favourable comparison with experimental results from GRP plates is presented.

INTRODUCTION

In general terms the effects of circular holes, or cut-outs, on the stability of thin rectangular plates has received little attention in the past for platework structures fabricated from conventional materials of construction. Consequently it is not surprising that the equivalent problem for composite plates has received even less attention. Bearing in mind the inevitability of access holes in a great many instances the problem clearly has considerable practical significance, particularly in the aerospace and shipbuilding industries.

After an exhaustive survey of previously published work on isotropic, e.g steel, thin, uniaxially compressed, plates with circular holes references [1] to [6] inclusive have come to light. These previous investigators have largely confined their thoughts to simply supported plates of square platform ($\lambda = 1$). A dearth of experimental evidence in this work is also noteworthy. Moreover, it is apparent that the diversity of mathematical techniques employed in these aforementioned studies have produced a considerable dichotomy of opinion in their published results. This is especially so in the case of large access holes, i.e. hole diameter/plate width ratios $\geqslant 0.5$. The issue is further confused by the considerable experimental scatter apparent when the limited published data is compared.

As yet no directly comparable studies on the equivalent problem for composite plates have been uncovered.

THEORETICAL ANALYSIS

Noting Fig. 1 the plate boundary conditions consistent with simply supported edges can be written :

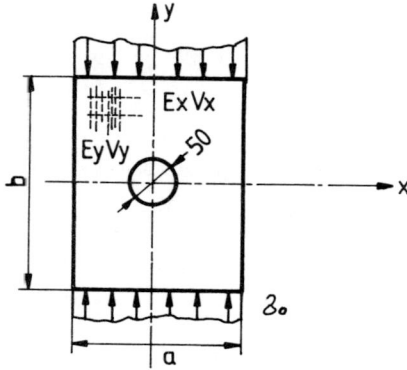

Fig. 1   Co-ordinate system.

at $x = \pm a/2$ at $y = \pm b/2$

$\omega = 0$ $\omega = 0$ zero edge
                                                                   lateral deflection

$$\frac{\partial^2 \omega}{\partial x^2} + \nu_y \frac{\partial^2 \omega}{\partial y^2} = 0 \qquad\qquad \frac{\partial^2 \omega}{\partial y^2} + \nu_x \frac{\partial^2 \omega}{\partial x^2} = 0 \qquad \text{zero edge moment} \tag{1}$$

Also, the boundary conditions at the periphery of a hole radius $r = r_o$ can be written :

$$(M_r)_{r=r_o} = 0 \qquad \text{and} \qquad \left(Q_r - \frac{\partial M_{r\theta}}{r \, \partial \theta}\right)_{r=r_o} = 0 \tag{2}$$

i.e. a stress-free hole boundary,

where

$$(M_r)_{r=r_o} = \frac{\partial^2 \omega}{\partial x^2} \left(D_x \cos^2\theta + D_y \nu_x \sin^2\theta\right)$$

$$+ \frac{\partial^2 \omega}{\partial y^2} \left(D_y \sin^2\theta + D_x \nu_y \cos^2\theta\right)$$

$$+ \frac{\partial^2 \omega}{\partial x \partial y} \left(\frac{t^3}{3} G \sin\theta \cos\theta\right) = 0$$

and
$$\left(Q_r - \frac{\partial \underset{r\theta}{M}}{r \; \partial \theta}\right)_{r=r_o} =$$

$$[\frac{\partial^3 \omega}{\partial x \partial y^2} (\frac{t^3}{6} * G + D_x \; \nu_y) \cos \theta + \frac{\partial^3 \omega}{\partial x^3} * D_x \cos \theta$$

$$+ \frac{\partial^3 \omega}{\partial y \partial x^2} (\frac{t^3}{6} G * D_y \; \nu_x) \sin \theta + \frac{\partial^3 \omega}{\partial y^3} * D_y \sin \theta]$$

$$-[\frac{\partial^2 \omega}{\partial x^2} (D_x \sin^2\theta + D_y \; \nu_x \sin^2\theta)$$

$$+ \frac{\partial^2 \omega}{\partial y^2} (D_y \sin^2\theta + D_x \; \nu_y \cos^2\theta)$$

$$+ \frac{\partial^2 \omega}{\partial x \partial y} (\frac{t^3}{3} G \sin \theta \cos \theta)] = 0$$

and
$$x = r_o \cos \theta$$

$$y = r_o \sin \theta$$

From a mathematical point of view the hole readily lends itself to the use of polar co-ordinates (r,θ) whereas the plate edges can be conveniently defined by cartesian co-ordinates (x,y).

However, it can be shown that the plate lateral deformations are closely approximated by

$$\omega = \omega_c \; (\cos \frac{\pi x}{a} \cos \frac{\pi y}{b} + Le^{-c \; (\frac{x^2}{a^2} + \frac{y^2}{b^2})}) \qquad (3)$$

In (3) the first term described the overall plate buckled shape with the second term taking account of the localised deformations in the vicinity of the hole.

Since the present work is not concerned with loads greater than the buckling load a simple energy solution can be employed.
When an orthotropic plate is subjected to uniaxial compression the total energy in the system can be written :

$$U = U_B - U_S \qquad (4)$$

where
$$U_B = \iint_A [\frac{D_x}{2} ((\frac{\partial^2 w}{\partial x^2})^2 + \nu_y (\frac{\partial^2 \omega}{\partial x^2} \frac{\partial^2 \omega}{\partial y^2})) + \frac{D_y}{2} ((\frac{\partial^2 w}{\partial y^2})^2$$

$$+ \nu_x (\frac{\partial^2 \omega}{\partial x^2}) (\frac{\partial^2 \omega}{\partial y^2})) + \frac{Gt^3}{6} (\frac{\partial^2 \omega}{\partial x \partial y})^2] \; dA \qquad (6)$$

and

$$U_S = \frac{\sigma_o}{2} \iint_A [\ \frac{\sigma_x}{\sigma_o}\ (\frac{\partial^2 \omega}{\partial x^2})^2 + \frac{2\tau_{xy}}{\sigma_o}\ (\frac{\partial^2 \omega}{\partial x^2}\ \frac{\partial^2 \omega}{\partial y^2}) + \frac{\sigma_y}{\sigma_o}\ (\frac{\partial^2 \omega}{\partial y^2})^2\ ]\ dA$$

The plate in-plane stresses can be adequately described by :

$$\sigma_x = \frac{\sigma_o}{2}\ \frac{r_o^2}{r^2}\ [\ ccs\ 2\theta + (2 - \frac{3r_o^2}{r^2})\ \cos 4\theta]$$

$$\sigma_y = \frac{\sigma_o}{2}\ [\ 2 + \frac{r_o^2}{r}\ (3 \cos 2\theta - (2 - \frac{3r_o^2}{r^2})\ \cos 4\theta)]$$

$$\tau_{xy} = \frac{\sigma_o}{2}\ \frac{r_o^2}{r^2}\ [-\sin 2\theta + (2 - \frac{3r_o^2}{r^2})\ \sin 4\theta] \qquad (7)$$

Noting the boundary conditions prevailing at the hole periphery (2), the constants C and L can be obtained for any given plate geometry, by conventional minimisation techniques. Thereafter substituting (3) and (7) into (4) and minimising the result according to minimum energy principles

i.e.

$$\frac{\partial U}{\partial \omega} = 0 \qquad (8)$$

a relationship between applied stress and plate deformations can be obtained.

EXPERIMENTAL INVESTIGATION

In view of the aforementioned lack of published experimental work on pierced rectangular orthotropic plates it was considered essential to initiate an extensive experimental programme to run in parallel with the theoretical analysis and corroborate its findings.

The test rig

A novel test rig capable of realistically applying uniaxial compression to rectangular planform plates was designed and its construction supervised. The salient details of this piece of equipment are given in Fig. 2. Essentially the rig applies uniaxial compressure loading in the vertical plane via a small hydraulic cylinder (A). The magnitude of this loading is accurately monitored by a pre-calibrated proving ring (B). A set of flat linear bearings (C) along with an adjustable bottom platten allows uniaxial compression to be accurately applied to a variety of plate aspect (length/breadth) ratios (λ). The rig is capable of simulating either simply supported or fully fixed plate edges by the use of interchangeable supports (D).

The test plates were vacuum laminated using unidirectional glass cloth embedded in a polyester resin matrix. After suitable post-curing the plates were accurately machined to a square planform of approximately 250 mm by 250 mm giving an aspect ratio $\lambda = 1.0$.

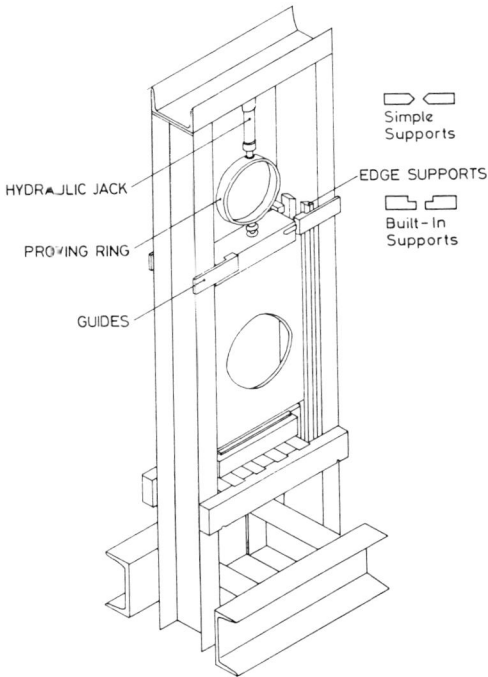

Each plate was tested in the unpierced condition to determine its critical
or buckling load. Thereafter hole diameters equal to 0.1, 0.2, 0.3, 0.4,
0.5, 0.6 and 0.7 of the plate width were machined and the respective plate
buckling loads determined.
At each of the aforementioned plate geometries a minimum of five tests were
carried out to check experimental repeatability. Thus the relationship between buckling load and hole diameter was experimentally determined for
square plates.

The test plates of thickness 1.9 mm possessed the following material
properties :

$$E_x = 3.1 \text{ GN/m}^2$$

$$E_y = 10.4 \text{ GN/m}^2$$

$$\nu_y = 0.300$$

$$G_{xy} = 2.142 \text{ GN/m}^2$$

COMPARISON BETWEEN THEORY AND EXPERIMENT

Isotropic pierced plates

In order to make comparison with the previous experiments cited in this paper, the present analysis was simplified for the case of plates with isotropic material properties. Figure 3 shows the present theoretical solution to favourably compare with past studies. It should be noted that the present analysis inherently considers plate edges subject to uniform stress loading and hence can be reasonably compared with references [1], [2] and [6]. However, references [3], [4] and [5] which consider uniformly displaced plate edges have also been included for completeness.

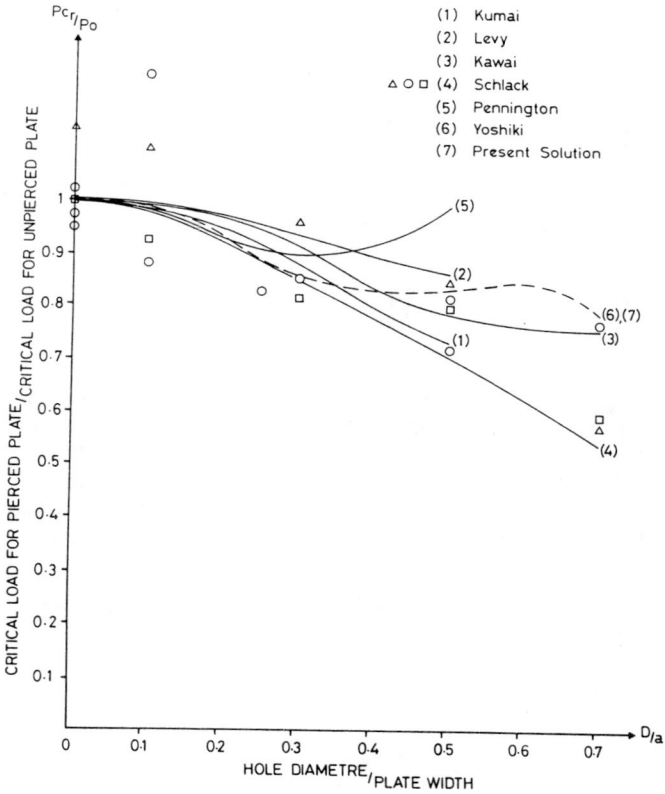

Composite pierced plates

Figure 4 compares the present theoretical analysis with tests carried out on GRP plates. In general, favourable comparison between theory and experiment is apparent within the range of plate geometries cited. Although at hole diameter to high plate width ratios this accuracy appears to diminish, it is suggested that this might well be a transition region between plate failure due to buckling and an alternative collapse mode. However, this extreme geometry is possibly outwidth practical interest.

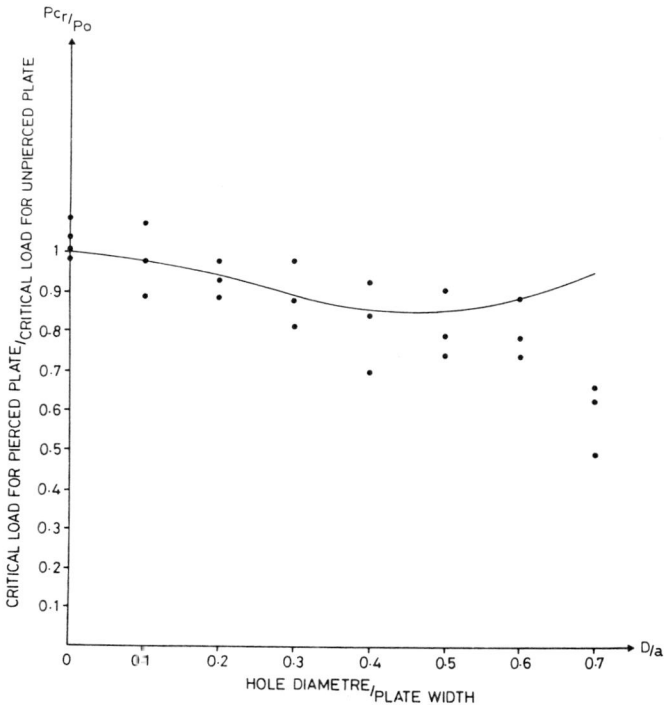

CONCLUDING REMARKS

The present theoretical analysis has been shown to compare favourably with earlier investigators and with a parallel experimental investigation. Although presently restricted to square plates it is fully capable of being extended to more general rectangular plate problems. This extension will be the subject of a future paper.

Acknowledgement

Part of this work was presented at the Reinforced Plastics Congress,
Brighton, U.K. November 1984.

REFERENCES

1. Kumai, T., Elastic stability of the square plate with a circular hole
   under edge thrust.
   Reports of the Research Institute for Applied Mechanics, vol. 1, no. 2,
   April 1952.

2. Levy, S., Woolley, R.M. and Kroll, W.A., Instability of a simply supported suare plate with a reinforced circular hole in edge compression.
   Journal of Research National Bureau of Standards, vol. 39, December 1947.

3. Kawai, T. and Ohtsubo, H., A method of solution for the complicated buckling problems of elastic plates with combined use of Rayleigh-Ritz"s
   procedure in the finite element method.
   Proceedings of the Second Air Force Conference on Matrix Methods in
   Structural Mechanics, Wright-Patterson AFB, October 1968.

4. Schlack, A.L., Experimental critical loads for perforated square plates.
   Proceedings of the Society for Experimental Stress Analysis, vol. 25,
   1968.

5. Pennington-Wann, W., Compressive buckling of perforated plate elements.
   Proceedings of the First Speciality Conference on Cold-formed Structures
   (1971), University of Missouri, Rolla, Missouri, April 1973, pp. 58-64.

6. Yoshiki, M., Fujita, Y., Kawamura, A. and Arai, H., Instability of plates
   with holes (First Report).
   Proceedings of the Society of Navel Architects of Japan, no. 122, December
   1967.

APPENDIX : NOMENCLATURE

a,b	plate dimensions
r_o	hole radius
r,θ	polar co-ordinates
t	plate thickness
x,y	cartesian co-ordinates
ω	plate lateral deflections
A	plate working area
D_x, D_y	flexural modulus for x and y directions respectively
G	shear modulus
L,c	constants dependant on hole geometry
ν_x, ν_y	Poisson's ratios for x and y directions respectively
σ_o	applied uniaxial compressive stress

LIST OF PARTICIPANTS

R. BARES
Czechoslovak Academy of Sciences, Czechoslovakia

C. BATHIAS
Université de Technologie de Compiège, France

F. BENEDIC
Aérospatiale, Suresnes, France

Y. BENVENISTE ★
School of Engineering, Tel-Aviv University
Ramat-Aviv, Tel-Aviv, Israel

H.F. BRINSON ★
Center for Adhesion Science, Department of Engineering Science and Mechanics
Virginia Polytechnic Institute and State University
Blacksburg, Virginia 24061, U.S.A.

R. BROUWER
Free University of Brussels (V.U.B.), Belgium

A.H. CARDON, chairman
Free University of Brussels
Dept. of Applied Continuum Mechanics, Faculty of Applied Sciences
B-1050 Brussels, Belgium

W.S. CARSWELL ★
National Engineering Laboratory
East Kilbride
Glasgow G13 3QN, Scotland.

R.M. CHRISTENSEN ★
Lawrence Livermore Laboratory
L-338 ; P.O. Box 808
Livermore, California 94550, U.S.A.

S. DE BECKER
Free University of Brussels, Belgium.

R. DECHAENE
State University of Ghent (R.U.G.), Belgium.

T. DE JONG ★
University of Technology Delft, Dept. of Aerospace Engineering
Klyverweg, 1
Delft, Netherlands.

P. DENIZET
C.E.A. - Centre d'Etudes de Bryères-le-Châtel, France.

244

W.P. DE WILDE
Free University of Brussels (V.U.B.), Belgium

W.W. FENG
Lawrence Livermore National Laboratory, California, U.S.A.

J.E. FITZGERALD ★
School of Civil Engineering, Georgia Institute of Technology
Atlanta, Georgia 30332, U.S.A.

J. FRELAT
Ecole Polytechnique, Palaiseau, France

G. FUHRMANN
Bundesanstalt für Materialprüfung (BAM), Berlin, W.-Germany

G. GAMSKI ★
University of Liège, Institut du Génie Civil
Quai Banning, 6
B-4000 Liège, Belgium

B. GUNS
State University of Ghent (R.U.G.), Belgium

P. HAMELIN ★
INSA - Service Bétons et Structures - 304
20, avenue A. Einstein
69621 Villeurbanne, France

Z. HASHIN ★
Tel Aviv University, Faculty of Engineering
Dept. of Solid Mechanics, Materials and Structures
Ramat-Aviv, 69978 Tel-Aviv, Israel

C.T. HERAKOVICH ★
Virginia Polytechnic Institute and State University
Dept. of Engineering Science and Mechanics
Blacksburg, Virginia 24061, U.S.A.

M.W. HYER ★
Virginia Polytechnic Institute and State University
Dept. of Engineering Science and Mechanics
Blacksburg, Virginia 24061, U.S.A.

K. KAMIMURA ★
Université de Compiègne (C.N.R.S.)
B.P. 233
F-60206 Compiègne Cedex, France

J.L. KING ★
University of Edingburgh, Dept. of Mechanical Engineering
King's Building, Mayfield Road
Edinburgh, EH9 3JL, Scotland

A. LESTIBOUDOIS
Free University of Brussels (V.U.B.), Belgium

K. LEVIN
FFA
Box 11021
S-16111 Bromma, Sweden

Ian M. MARSHALL ★
Paisley College of Technology
High street
Paisley, PA1 2BE, Scotland

P. MEIJERS
Delft University of Technology, Delft, Netherlands

B. NARMON
Free University of Brussels (V.U.B.), Belgium

J. PABIOT
Centre d'Etude des Matières Plastiques, Paris, France

G. PATFOORT
Free University of Brussels (V.U.B.), Belgium

F. PEREZ
Aérospatiale, Les Mureaux Cedex, France

D. PHAN VAN
Dow Chemical Netherlands, Terneuzen, Netherlands

J.C. RADON ★
Imperial College, Dept. of Mechanical Engineering
Exhibition Road
London SW7 2AZ, England

A. RÜBBEN
RWTH Aachen, Germany

K. SCHULTE ★
DFVLR - Institut für Werkstoff-Forschung
Postfach 906058
D-5000 Köln 90, Germany

F. SIDOROFF ★
Ecole Centrale de Lyon
B.P. 163
F-69131 Ecully, France

H. SOL
Free University of Brussels (V.U.B.), Belgium

K.K.U. STELLBRINK ★
DFVLR. WB-BK
Pfaffenwaldring, 38-40
D-7000 Stuttgart 80, Germany

P.S. THEOCARIS ★
Athens National Technical University
5, Heroes of Polytechnion Avenue
GR-157 73 Athens, Greece.

M.E. TUTTLE
RISØ National Laboratory, Roskilde, Denmark

D. VALENTIN ★
Ecole des Mines de Paris
B.P. 87
F-91003 Evry Cedex, France

W. VAN DEN BRANDE
Free University of Brussels (V.U.B.), Belgium

D. VAN DER SANDEN
Free University of Brussels (V.U.B.), Belgium

G.D. VAN DER WEIJDE
Fokker BV, Schiphol, Netherlands

R. VAN GEEN
Free University of Brussels (V.U.B.), Belgium

D. VAN GEMERT ★
Catholic University of Leuven, Dept. of Construction
de Croylaan, 2
B-3030 Heverlee, Belgium

A. VAUTRAIN ★
Ecole des Mines de St.-Etienne
158 Cours Fauriel
F-42023 Saint-Etienne Cedex, France

G. VERCHERY, chairman
Ecole des Mines de St.-Etienne
158 Cours Fauriel
F-420234 Saint-Etienne Cedex, France

J. VEREECKEN
Free University of Brussels (V.U.B.), Belgium

I. VERPOEST
Catholic University of Leuven, Belgium

F. VIDOUSE
S.N.E.A., Artix, France

J. WASTIELS
Free University of Brussels (V.U.B.), Belgium

M. WEVERS ★
Catholic University of Leuven, Dept. of Metallurgy
de Croylaan, 2
B-3030 Heverlee, Belgium

Yong-Sok O ★
Dow Pipe Systems
F-76410 Tourville La Rivière, France

INDEX OF CONTRIBUTORS

SUBJECT INDEX

252